MNEMONICS FOR STUDY

with English-Spanish glossary

Also by Fiona McPherson

Indo-European Cognate Dictionary

Mnemonics for Study (2nd ed.)

Mnemonics for Study (2nd ed.): Italian edition

My Memory Journal

Successful Learning Simplified: A Visual Guide

How to Approach Learning: What teachers and students should know about succeeding in school

How to Learn: The 10 principles of effective practice and revision

Effective Notetaking (2nd ed.)

Planning to Remember: How to remember what you're doing and what you plan to do

Perfect Memory Training

The Memory Key

Mnemonics for Study

(2ND ED.)

WITH
ENGLISH-SPANISH GLOSSARY

Fiona McPherson, PhD

Wayz Press
Wellington, New Zealand

Published 2018 by Wayz Press, Wellington, New Zealand.

Copyright © 2018 by Fiona McPherson.

All rights reserved.

No part of this publication may be reproduced, stored in a retrieval system, or transmitted in any form or by any means, electronic, mechanical, recording or otherwise, without the prior written permission of Wayz Press, a subsidiary of Capital Research Limited.

ISBN 978-1-927166-46-8

To report errors, please email errata@wayz.co.nz

For additional resources and up-to-date information about any errors, go to the Mempowered website at www.mempowered.com

Contents

1. INTRODUCTION TO MNEMONICS	**1**
What are mnemonics and what are they good for?	1
Glossary	3
Why are mnemonics effective?	4
Glossary	8
About imagery	10
Individual differences	11
Glossary	12
Using imagery	13
Glossary	15
Review	16
PART I SIMPLE VERBAL MNEMONICS	**18**
2. FIRST LETTER MNEMONICS	**19**
The two types of first-letter mnemonics	19
Glossary	21
How to create effective first-letter mnemonics	23
Glossary	27
Problems with first-letter mnemonics	28
Glossary	30

When first-letter mnemonics are a good strategy to use, and when they're not	31
Glossary	34
Review	35

3. RHYTHM & RHYME — 37

Some familiar mnemonic jingles	37
Glossary	39
Singing to remember	40
Glossary	42
Spoken rhythm	44
Glossary	47
Review	48

PART II KEYWORD STRATEGIES — 49

4. THE KEYWORD METHOD — 51

Glossary	52
Some examples to practice	52
Glossary	59
Creating good keywords	61
Glossary	69
How effective is the keyword method?	70
Glossary	72
Limitations of the keyword method	73
Glossary	74

Remembering for the long term	75
Glossary	77
Comparing the keyword mnemonic to other strategies	78
Glossary	80
Tasks for which the keyword method is useful	80
Using the keyword mnemonic to remember gender	81
Non-European languages	81
Glossary	82
Review	83
5. EXTENSIONS OF THE KEYWORD METHOD	**86**
More than words	86
Glossary	87
Applying the keyword method to text	88
Glossary	93
The face-name mnemonic	95
Applying the face-name mnemonic to art & artists	95
Applying the face-name mnemonic to animals	97
Extending the mnemonic to taxonomic & attribute information	98
Glossary	100
Review	102
PART III LIST MNEMONICS	**103**
Glossary	105

6. THE STORY METHOD — 107

- Examples — 107
  - Remembering word lists — 107
  - Remembering text — 110
- Pros & cons of the story method — 113
  - Glossary — 115
- Review — 117

7. THE PLACE METHOD — 119

- Using the place method — 120
  - Some advice from antiquity — 123
- When to use the place method — 125
  - Glossary — 127
- Review — 128

8. THE PEGWORD MNEMONIC — 130

- Applying the pegword method — 134
  - Glossary — 135
- Review — 136

9. THE LINK METHOD — 138

- Effectiveness of the link method — 140
  - Glossary — 141
- Review — 142

PART IV ADVANCED MNEMONICS — 143

10. CODING MNEMONICS   144

A system for remembering numbers   144

Extending the coding mnemonic with other mnemonics   149

Practical uses for coding mnemonics   151

Dealing with decimals   158

Retrieval   159

Other languages may work better!   159

Using the coding system to extend the pegword mnemonic   160

Glossary   162

Review   164

11. MASTERING MNEMONICS   169

What mnemonics are good for   169

Assessing the text and the task   169

Choosing the right strategy for the task   172

Choosing the strategies that are right for you   175

Successful strategies need practice   176

Glossary   179

Review   181

Summary of mnemonic strategies   183

12. CASE STUDY   185

The Geological Time Scale   186

Memorizing new words, lists and dates	187
Acronyms	187
Looking for meaning	189
Glossary	194
Chunking your information	195
Using a Coding Mnemonic to remember dates	198
Glossary	199
More chunking	199
Visuals always help	203
Glossary	206
One era at a time	207
Paleozoic Era	207
Glossary	213
Time to review	214
Review 12.1	219
Mesozoic Era	221
Review 12.2	224
Glossary	227
Cenozoic Era	228
Review 12.3	234
Review 12.4	236
Glossary	239

	Precambrian	240
	The big picture	245
	Review 12.5	248
	Glossary	252
	Mnemonic pictures for practice	253

Answers — 255
References — 261

	Introduction to mnemonics	261
	First-letter mnemonics	262
	Rhythm & Rhyme	262
	Keyword method	263
	Extensions to the keyword method	270
	Story method	272
	Place method	273
	Pegword method	274
	Link method	274
	Coding mnemonic	275
	Mastering mnemonics	276

English-Spanish Glossary (Alphabetical) — 278

	Acrostics & acronyms	312

Note for this Spanish edition

This version of *Mnemonics for Study* has an extensive English-Spanish glossary to assist Spanish-language readers. The relevant glossary is provided after each section, and these section glossaries are all included in the Table of Contents for easy reference. A complete glossary in alphabetical order is also included at the end of the book.

Introduction to mnemonics

What are mnemonics and what are they good for?

Any memory improving strategy can, of course, be termed a mnemonic strategy, but in its more specific meaning, mnemonic refers to artificial memory aids such as stories, rhymes, acronyms, and more complex strategies involving verbal mediators or visual imagery, such as the journey method or method of loci, the pegword method, and the keyword method.

We will get to each of these in due course, but first we need to consider the benefits and limitations of such mnemonics, and in particular when you should use them in the course of study and when you should not.

The most important thing to understand is that mnemonics do not help you understand your material. They do not help with comprehension; they do not help you make meaningful connections.

The purpose of mnemonics is simply to help you remember

something — not by understanding it, not by incorporating it into your developing "expert database", but simply in the manner of a parrot. They are used to enable you to regurgitate information.

That sounds terribly contemptuous, but if I considered there was no value in mnemonics I wouldn't be devoting this book to them. The ability to regurgitate information on demand is undeniably a useful one — indeed, in the context of examinations, often a vital one!

Even in the context of material you need to understand, there are often details that must simply be memorized — names of things, technical words, lists of principles, and so on. Moreover, mnemonics can help you remember tags or labels that allow you to access clusters of meaningful information — for example, headings of a speech or main points for exam essays.

For both these reasons, mnemonics are a valuable assistance to building up expertise in a subject, as well as in helping you 'cram' for an exam.

However, despite a number of studies showing the effectiveness of mnemonic strategies, these remain the least frequently used formal memory aid used by students[1]. Perhaps the main reason for this is that their effectiveness is not intuitively obvious — truly, no one really believes that these 'tricks' can so remarkably improve memory until they try them for themselves.

But I can help you believe (and belief is vitally necessary if you're going to make the effort to use, and keep on using, any memory strategy) if I explain why they work. It's also important to understand the principles involved if you're going to fully master these techniques — by which I mean, know when and when not to use them, and how to use them flexibly.

Do note that this is a very bare-bones account of the principles.

For more details, I refer you to my book *The Memory Key*, available as a digital download, or to the revised edition put out in paperback by Random House as *Perfect Memory Training*.

Glossary

memory improving strategy — estrategia para mejorar la memoria

mnemonic strategy — estrategia mnemotécnica

artificial memory aids — ayudas artificiales de memoria

stories — cuentos

rhymes — rimas

acronyms — acrónimos

verbal — verbal, para hacer con palabras

mediators — mediadores, intermediarios

imagery — imágenes

journey — viaje

method of loci — método de loci

pegword method — método della parola aggancio

keyword — palabras clave

benefits — beneficios

limitations — limitaciones

comprehension — comprensión

meaningful connections — conexiones significativas

incorporating — incorporando

expert database — base de datos de expertos

regurgitate — regurgitar (repetir peatonalmente)
information — información
context — contexto
examinations — exámenes
memorized — memorizado
technical words — términos técnicos
clusters — racimos, grupos
expertise — pericia
subject — tema
cram for an exam — prepararse apresuradamente
effectiveness — eficacia
intuitively — intuitivamente
obvious — obvio
tricks — trucos
belief — creencia
strategy — estrategia
master — dominar
techniques — técnicas
flexibly — de una manera flexible
bare-bones — lo esencial

Why are mnemonics effective?

Let's think about the basic principles of how memory works. The strength of memory codes, and thus the ease with which they can be found, is a function largely of repetition. (For those who haven't read *The Memory Key*, let me note that I habitually refer to information encoded in memory as memory codes to emphasize that memories are not faithful and complete recordings, but highly selected and edited.) Quite simply, the more often you experience something (a word, an event, a person, anything), the stronger and more easily recalled your memory for that thing will be.

This is why the most basic memory strategy — the simplest, and the first learned — is rote repetition.

Repetition is how we hold items in working memory, that is, "in mind". When we are told a phone number and have to remember it long enough to either dial it or write it down, most of us repeat it frantically. This is because we can only hold something in working memory by keeping it active, and this is the simplest way of doing so.

Spaced repetition — repetition at intervals of time — is how we cement most of our memory codes in our long-term memory store. If you make no deliberate attempt to learn a phone number, yet use it often, you will inevitably come to know it (although how many repetitions that will take depends on several factors, including individual variability).

But most of us come to realize that repetition is not, on its own, the most effective strategy for learning, and when we deliberately wish to learn something, we generally incorporate other, more elaborative, strategies.

Why do we do that? If memory codes are strengthened by repetition, why isn't it enough to simply repeat?

Well, it is. Repetition IS enough. But it's boring. That's point one.

Point two is that making memory codes more easily found

(which is after all the point of the exercise) is not solely achieved by making the memory codes stronger. Also important is making lots of connections. Memory codes are held in a network. We find a particular one by following a trail of linked codes. Clearly, the more trails that lead to the code you're looking for, the more likely you are to find it.

Elaborative strategies — mnemonic strategies, organizational strategies — work on this aspect. They are designed to increase the number of links (connections) a memory code has, and therefore the number of different routes you can take to it.

Thus, when we note that a lamprey is an "eel-like aquatic vertebrate with sucker mouth", we will probably make links with eels, with fish, with the sea. If we recall that Henry I was said to have died from a surfeit of lampreys, we have made another link. Which in turn might bring in yet another link, that Ngaio Marsh once wrote a mystery entitled "A surfeit of lampreys". And if you've read the book, this will be a good link, being itself rich in links. (As the earlier link would be if you happen to be knowledgeable about Henry I).

On the other hand, in the absence of any knowledge about lampreys, you could have made a mnemonic link with the word "lamp", and imagined an eel-like fish with lamps in its eyes.

So, both types of elaborative strategy have the same goal — to increase the number of connections.

But mnemonic links are weaker in the sense that they are arbitrary. Their value comes in those circumstances when either you lack the knowledge to make meaningful connections, or there is in fact no meaningful connection to be made (this is why mnemonics are so popular for vocabulary learning, and for the learning of lists and other ordered information).

Mnemonic strategies have therefore had particular success in the learning of other languages. However, if you can make a meaningful connection, that will be more effective.

For example, in Spanish the word *surgir* means to appear, arise. If you connect this to the word *surge*, from the Latin *surgere*, to rise, then you have a meaningful connection, and you won't, it is clear, have much trouble when you come across the word. However, if your English vocabulary does not include the word *surge*, you might make instead a mnemonic connection, such as *surgir* sounds like *sugar*, so you make a mental image involving rising sugar.

Now, consider each of these situations. Say you don't come across the word again for a month. When you do, which of these connections is more likely to bring forth the correct meaning?

But of course, it is not always possible to make meaningful connections, and this is where mnemonics are so useful.

Additionally, sequence is often not obviously meaningful (although it may become so when you have a deeper understanding of the subject), and mnemonics are particularly good for ordered information.

The thing to remember however, is that you haven't overcome the need for repetition. These strategies are adjuncts. The basic principles must always be remembered:

- Memory codes are made stronger by repetition.
- Links are made stronger by repetition.
- If you don't practice the mnemonic, it won't be remembered.

The same is true for any connection, but meaningful connections are inherently stronger, so they don't need as many repetitions.

Points to Remember

Memory codes are made stronger by repetition.

Memory codes are made easier to find by increasing the number of links they have to other memory codes.

Elaborative strategies make connections with existing codes.

Some elaborative strategies make meaningful connections between memory codes — these are stronger.

Mnemonic strategies make connections that are not meaningful.

Mnemonic strategies are most useful:

- where there are no meaningful connections to be made, or you lack the knowledge to make meaningful connections
- where you need to remember items in sequence.

Glossary

basic principles — principios básicos

memory codes — códigos de memoria

function — función

repetition — repetición

habitually — por costumbre

encoded — codificado

faithful — fiel

recordings — grabaciones

selected — seleccionado

rote repetition — repetición mecánica

working memory — memoria de trabajo

in mind — en mente

repeat — repetir

active — activo

spaced repetition — repetición espaciada

intervals — intervalos

cement — cimentamos, consolidamos

long-term memory — memoria a largo plazo

store — base de datos

deliberate attempt — intento deliberado

factors — factores

individual variability — variabilidad individual

strengthened — fortalecido

boring — aburrido

connections — conexiones

network — red

organizational — organizativo

aspect — aspecto

links — enlaces, conexiones

routes — caminos

lamprey — lamprea

aquatic — acuático

vertebrate — vertebrado

sucker — ventosa
surfeit — exceso
knowledgeable — bien informado
lamp — lámpara
imagined — imaginado
weaker — más débil
arbitrary — arbitrario
circumstances — circunstancias
vocabulary — vocabulario
learning — aprendizaje
lists — listas
ordered — ordenado
situations — situaciones
sequence — secuencia
adjuncts — adjuntos, usado en conjunto con otras estrategias
remembered — recordado

About imagery

The more complex mnemonic strategies are usually based on visual images. This causes people who feel that their ability to 'see' mental images is poor, to think that mnemonics are of no use to them. That would be overly hasty.

Although imagery is certainly an effective tool, there is nothing particularly special about it. The big advantage of imagery is that it provides an easy way of connecting information that is not otherwise readily connected. However, providing verbal links can be equally effective.

Individual differences

Moreover, although there is undoubtedly considerable variation between people in terms of their abilities to visualize images, only a very small percentage of people don't visualize. A similarly small percentage make extremely vivid images. Somewhere in between are the rest of us.

My own feeling is that many people don't realize the extent to which they form visual images. You don't need clear television-quality mental images to visualize usefully! When you're reading a novel, for example, you may well have no conscious awareness of the pictures being created in your mind, but if you see a movie adaptation of the book you'll immediately notice all the visual images that are 'wrong' (such as what the hero looks like).

If you're uncertain about your visualization abilities, you might find these signs interesting:

High visualizers are more easily tricked into thinking imagined experiences have really happened — they create false memories more easily. So if you know you're prone to that, that's a sign that you form good images!

There's also some evidence$_2$ that high visualizers are better at fighting the Stroop effect. The Stroop effect concerns color-name interference: when you see the name of a color written in the same color, that's easier to process than when the color doesn't match the name (for example, 'red' written in blue ink). Comparison of the different reaction times has been used to test attention, executive function, and processing speed, and less directly the presence of various disorders.

More recently a study that looked at differences in brain

activity as people imagined a visual scene, found that not only were there noticeable differences between good and poor visualizers (good visualizers had more activity in their visual cortex), but that this correlated with performance on the Stroop test. Those showing greater activity in their visual cortex (good visualizers) were slower at naming colors when the names matched the color they were written in (this is of course the opposite of what is usually found). This is a very small study (8 people) and we can't draw too firm a conclusion, but it is interesting. There are places on the web where you can get a feel for your performance on this task.

But the most important thing to note is that visualization is a skill that is strengthened with practice. If you persevere with trying to make good visual images, you will get better at it.

Glossary

images — imágenes mentales

variation — variación

abilities — habilidades

visualize — visualizar, imaginar

conscious awareness — conocimiento consciente

visualizers — visualizadores

experiences — experiencias

prone — propenso a

interference — interferencia

match — corresponde a

comparison — comparación

different — diferente

reaction times — tiempos de reacción

attention — atención

executive function — función ejecutiva, procesos cognitivos que lo ayudan a controlar sus pensamientos y comportamiento

processing speed — velocidad de procesamiento

visual scene — escena visual

noticeable — que se nota, evidente

visual cortex — corteza visual

correlated — correlacionado

performance — ejecución

practice — práctica

persevere — perseverar

Using imagery

For myself, I am very verbal (not surprising in a writer!), but that doesn't mean images, albeit not particularly vivid ones, aren't being formed in my mind. What I find works best for me, and probably what will work best for most of you, if you're not at the extreme ends of visualization ability, is the use of both words and images.

That is, when you're using a verbal mnemonic, aim for words that are easy to visualize; when you're using a visual mnemonic, make sure the pictures have associated word labels, and keep them in mind while you're imagining the picture.

Most mnemonic strategies, despite being usually described as 'visual' or 'verbal', do in fact combine both aspects, and you may emphasize the visual or the verbal aspect as suits you.

It is usually emphasized that bizarre images are remembered much better, but there is no clear evidence for this. Indeed in many studies ordinary images are remembered slightly better. One of the problems is that people usually find it harder to create bizarre images. Unless you have a natural talent for thinking up bizarre images, it is probably not worth bothering about.

Whether bizarre or not, images generally take longer to construct than verbal phrases. If you want to construct them 'on the fly', as you hear information in conversation or in a presentation like a lecture, then you need to have practiced the skill a great deal.

The critical point to remember is that images, and words, work well as mnemonics only to the extent that they are bound together. Thus, an image needs to be interactive — tying the bits of information tightly together.

Don't worry if you're not sure exactly what this means! It will become clear as we discuss the various mnemonic techniques. For now, I just ask you to bear it in mind.

Points to Remember

Images are effective to the extent that they link information.

Images are not inherently superior to words.

Bizarre images are not necessarily better recalled than common ones.

Effective images have the elements interacting with each other.

Glossary

albeit — aunque

extreme — extremos

word labels — etiquetas de palabras

emphasized — enfatizado

bizarre — extraño

ordinary — ordinario

talent — talento, dotes

construct — construir

phrases — frases

on the fly — sobre la marcha, improvisado

conversation — conversacion

presentation — presentación

lecture — conferencia

skill — técnica, habilidad

critical — crítico

interactive — interactivo

inherently — intrínsecamente

superior — mayor

necessarily — necesariamente

recalled — recordado

common — común, ordinario

elements — elementos

Review 1.1

1. What makes memories easy to remember?

   a) repetition

   b) multiple connections

   c) vividness

   d) strangeness

   e) strong connections

2. Which of these are mnemonic strategies (in the narrower sense of the word)?

   a) acrostics

   b) rhymes

   c) mind maps

   d) graphic organizers

   e) keyword method

3. Which is better to help you remember:

   a) visual images

   b) verbal phrases

   c) it depends

4. Which of these strategies are better to help you understand meaningful information?

   a) keyword method

   b) stories

c) concept maps

d) taking notes

5. When are mnemonics a good idea?

   a) when the information isn't connected in a meaningful way

   b) when the information is densely connected

   c) when the information consists of familiar items in a specific order

   d) when the information consists of brief items

   e) when you want to develop expertise in the subject

6. You're either born with the ability to make good visual images, or you're not. T / F

7. The most important thing about mnemonic images is:

   a) being brightly colored

   b) being large

   c) being simple

   d) being interesting

   e) having interactive elements

8. Why are mnemonics effective for arbitrary information?

   a) they are bizarre

   b) they make unusual pictures

   c) they make connections where there aren't 'natural' ones

   d) they add interest to boring information

Part I

Simple Verbal Mnemonics

In this section, we're going to explore some simple verbal mnemonics that you will already be familiar with — acronyms, acrostics, rhymes, and songs.

First-letter mnemonics

The two types of first-letter mnemonics

First-letter mnemonics are, as their name suggests, memory strategies that use the initial letters of words as aids to remembering. This can be an effective technique because initial letters are helpful retrieval cues, as anyone who has endeavored to remember something by mentally running through the letters of the alphabet can attest to.

There are two types of first-letter mnemonic:

- **acronyms:** initial letters form a meaningful word;
- **acrostics:** initial letters are used as the initial letters of other words to make a meaningful phrase

ROY G. BIV is an acronym (for the colors of the rainbow), and **R**ichard **O**f **Y**ork **G**ives **B**attle **I**n **V**ain is an acrostic for the same information. Similarly, the acronym **FACE** is used to remember the notes in the spaces of the treble staff, and the acrostic **E**very **G**ood **B**oy **D**eserves **F**ruit for the notes on the lines of the treble staff.

Here's some more well-known ones. Some acronyms first:

MRS GREN — the characteristics of living things: **M**ovement, **R**espiration, **S**ensitivity, **G**rowth, **R**eproduction, **E**xcretion, **N**utrition.

BEDMAS — the order of mathematical operations: **B**rackets, **E**xponent, **D**ivision, **M**ultiplication, **A**ddition, **S**ubtraction.

HOMES — the Great Lakes in the U.S.A.: **H**uron, **O**ntario, **M**ichigan, **E**rie, **S**uperior.

And some acrostics:

My **V**ery **E**ager **M**other **J**ust **S**erved **U**s **N**ine **P**izzas — the order of the planets: **M**ercury, **V**enus, **E**arth, **M**ars, **J**upiter, **S**aturn, **U**ranus, **N**eptune, **P**luto.

Father **C**harles **G**oes **D**own **A**nd **E**nds **B**attle — the order of sharps in music

King **P**hillip **C**ame **O**ver **F**rom **G**reat **S**pain — the order of categories in the taxonomy of living things: **K**ingdom, **P**hylum, **C**lass, **O**rder, **F**amily, **G**enus, **S**pecies

It's likely that you'll know very different acrostics for these same items. That's one difference between acronyms and acrostics — the same acronyms are likely to be known to everyone, but acrostics are much more varied. The reason's not hard to seek — clearly there are infinite possibilities for acrostics, but very very limited possibilities for acronyms.

This means, of course, that opportunities to use acronyms are also very limited. It is only rarely that the initial letters of a group of items you wish to learn will form a word or series of words or at least a pseudo-word (a series of letters that do not form a word but are pronounceable as one — like BEDMAS).

Nothing is going to make MVEMJSUNP (the order of planets) memorable in itself, even if you break it up into vaguely intelligible bits, like this: M.V. Em J. Sun P. (although that does help — say it and you'll see why). Acrostics, on the other hand, are easy to create, and any string of items can be expressed in that form. For example:

My **V**ery **E**arnest **M**other **J**umped **S**even **U**mbrellas **N**ear **P**aris

Men **V**iew **E**nemies **M**ildly **J**uiced **S**ince **U**nited **N**ations **P**arty

Michael **V**oted **E**very **M**ay **J**udiciously **S**ince **U**nion **N**ewsletters **P**lunged

I created those as I typed; it's not difficult. But of course the aim is not simply to devise acrostics — it's to create *good* acrostics. That is, memorable ones. And that is not quite as easy.

Glossary

planets — planetas

initial — inicial

retrieval — recuperación

cues — señales

endeavored — esforzado

attest — dar fe, testificar

notes — notas

spaces — espacios
treble staff — pentagrama clave del sol
characteristic — característica
mathematical operations — operaciones matemáticas
order — orden
sharps — sostenidos
categories — categorías
taxonomy — taxonomía
opportunities — oportunidades
pronounceable — pronunciable
intelligible — inteligible
string — cadena, secuencia
devise — idear, inventar
sentence — oración, frase

Mercury, Venus, Earth, Mars, Jupiter, Saturn, Uranus, Neptune, Pluto — Mercurio, Venus, Tierra, Marte, Júpiter, Saturno, Urano, Neptuno, Plutón

Kingdom, Phylum, Class, Order, Family, Genus, Species — Reino, Filo, Clase, Orden, Familia, Género, Especie

Movement, Respiration, Sensitivity, Growth, Reproduction, Excretion, Nutrition — Movimiento, Respiración, Sensibilidad, Crecimiento, Reproducción, Excreción, Nutrición

Brackets, Exponent, Division, Multiplication, Addition, Subtraction — Soportes, Exponente, División, Multiplicación, Adición, Sustracción

How to create effective first-letter mnemonics

Let's start with one of the best-known mnemonics in geography, in the United States at least: HOMES — an acronym for the Great Lakes. Less well-known are sentence mnemonics to help you remember the geographical order of the lakes (from west to east or east to west), or their relative size. There are a few around. Here's some I made up:

Simon **M**akes **H**erons **E**at **O**lives (the Great Lakes from west to east)

Oliver **E**ats **H**errings **M**arinated **S**lowly (from east to west)

Simon **H**as **M**any **E**legant **O**wls (in order of size)

Now, these are all useful mnemonics, but they are only useful in very particular circumstances, glaringly obvious to non-North Americans at least. Namely, you need to already know the names of the Great Lakes.

All the lakes have fairly obvious cues: Superior is a familiar word; Erie is very close to the word eerie; Huron is very close to heron; both Ontario and Michigan are names for the province/state they're in. So if you aren't that familiar with the names of the lakes, rather than HOMES, you would be better with an acrostic like this one:

A heron is superior in Ontario but eerie in Michigan.

In fact, because HOMES is such a good acronym, being a short, very familiar word, the best mnemonic would be:

A heron is superior in Ontario but eerie in Michigan HOMES.

This sentence makes sense, and ties the two mnemonics together. This is good because both have value, and have slightly different functions. HOMES is much easier to remember and provides valuable first-letter retrieval cues; the acrostic provides more detailed cues for the items.

But neither the acrostic nor the acronym provides any order information. Let's look at my suggested acrostics for order:

Simon **M**akes **H**erons **E**at **O**lives (the Great Lakes from west to east)

Oliver **E**ats **H**errings **M**arinated **S**lowly (from east to west)

Simon **H**as **M**any **E**legant **O**wls (in order of size)

There are two obvious problems with these:

- they provide no information to help you with remembering the items themselves other than the first-letter retrieval cues, and
- they provide no clues to tell you what particular order is being specified.

Here are some examples that might be better for the size order:

Simon **H**as **M**any **E**normous **O**wls

Superior **H**erons **M**ight **E**at **O**wls

Superior **H**erons **M**ight **E**at **O**reos

Sizing **H**erons **M**ight **E**fface **O**wls

These words are better reminders of the items, for the most part. "Owls" is not a good cue for Ontario, but unfortunately (though not uncommonly), there are few words reminiscent of

Ontario! "Oreos" is probably a better one, but only for those who are familiar with Oreos (personally, not being an American, I know of them only by repute — which isn't really enough to make them a good cue for me).

"Superior" might be a good enough cue for you to put the acrostic in its proper context, because superior does vaguely have connotations of size. But it may not — hence the suggested use of "Sizing" instead of "Superior".

On the other hand, "Sizing" is not a particularly good cue for "Superior". So which of these words would be more effective for you depends on whether it's more important for you to have a clue to the name or the function of the acrostic.

The clue to context doesn't have to be in the first word (in the first example, the fourth word, "Enormous", is the clue), but I suspect it's a good idea, where possible. Here are some examples for direction order that incorporate hints that direction provides the order:

From east to west:

Oriental **E**nemies **H**ave **M**arine **S**nails

Oriental **E**ast **H**eads **M**ore **S**unset

The first of these makes a little more sense, but only the first word provides a clue to the context (east to west). The second is full of clues that this is about direction and the direction is east to west, but doesn't really make sense as a sentence. Neither provide any clues to the items themselves; they are there only to provide first-letter retrieval cues.

Maybe we'll have more luck with the opposite direction (you only need to know either west or east, *or* east to west, after all!).

From west to east:

Sunset Moves Heavily East Orientally

Sunset Moves Heavily East [to the] Orient

Simon May Head East Occasionally

These examples confirm what was suggested in the earlier examples — you can't provide both direction clues and content clues; the words can't bear the double burden. You have to choose which is more important to you. Or, of course, learn two separate mnemonics (or indeed, three, if size is also important to you). There's no particular problem with that, as long as the mnemonics provide the needed cues as to context.

These last two acrostics also provide examples of two permissible actions: making up words (orientally), and putting in small words that don't count (i.e., they're not to be considered when pulling out the first-letter cues; they're only there to help the acrostic make sense).

I say these actions are permissible; I don't say they're desirable. Both should only be resorted to when better alternatives fail, for invented words are less memorable, and redundant words are potentially confusing.

The last example I give makes more sense and doesn't have these drawbacks; it doesn't, however, have the context clue in the first word. On the other hand, I think the sentence as a whole provides a strong enough context clue that this doesn't matter (the point being, that a coherent sentence will be treated more readily as a whole, rather than processed word by word).

This analysis has, I hoped, suggested a number of rules for creating effective first-letter mnemonics, but before we summarize these, we need to consider some problems with this type of mnemonic.

Glossary

glaringly — claramente obvio
valuable — valioso
detailed — detallado
suggested — sugirió
clues — pistas
specified — especificado
examples — ejemplos
reminders — recordatorios
items — artículos
unfortunately — desafortunadamente
not uncommonly — no es raro
reminiscent — recuerdan
by repute — por su reputación
connotations — connotaciones
direction — dirección
opposite — opuesto
confirm — confirman
double burden — doble carga
permissible — permisibles, admisibles
actions — acciones
making up — componiendo, inventando
considered — considerado
desirable — deseables, aconsejables

resorted to — recurrido a
invented — inventadas
redundant — redundantes
potentially — potencialmente
confusing — confuso
drawbacks — desventajas
coherent — coherente, lógico
readily — fácilmente
processed — procesarse
summarize — resumir
problems — problemas

Problems with first-letter mnemonics

Medical students are probably the group who use first-letter mnemonics most. Here's a medical example that demonstrates a common problem with first-letter mnemonics:

> On Old Olympia's Towering Top A Finn And German Vault And Hop

This is a mnemonic for remembering the cranial nerves: olfactory, optic, oculomotor, trochlear, trigeminal, abducens, facial, auditory, glossopharyngeal, vagus, accessory, and hypoglossal. Of course, reiterating my earlier point, the mnemonic wouldn't help most of us remember this information, because we don't know these names. But there's another problem with this acrostic: three Os, two Ts and three As.

This is a particular problem when the purpose of the acrostic is to remind you of the precise order of items, for obvious reasons. In such a case, you need to use words that distinguish between similar items. Thus, a better acrostic for our medical students might be:

Oliver **Op**erates **Oc**casional **Tro**pical **Tri**cks **Ab**surdly For **Au**stralian **G**ymnasts **V**aulting **Ac**tual **H**elicopters

Except that the traditional acrostic does have two big advantages that make it a much more memorable sentence: rhythm and rhyme. Say them both aloud, and you'll see what I mean.

Let's try for an acrostic that contains the vital information *and* is memorable.

Oliver **Op**ens **Oc**eans; **Tro**pical **Tri**ps **Ab**et; **F**abulous **Au**thors **G**ushing; **V**iolent **Ac**ts **H**inted

Okay, this isn't very good either, and it took a little while to come up with. I've tried to distinguish the same-initial terms by including the second letter. The problem is, this additional constraint makes a big difference in limiting the possibilities.

Also, of course, creating an acrostic with rhyme and rhythm requires a great deal more work than simply creating a meaningful sentence. Rhyme and rhythm do, however, render the acrostic considerably more memorable.

In fact, were I trying to memorize the 12 cranial nerves, I wouldn't use a first-letter mnemonic. Let us consider what you need to learn:

- the names of each nerve
- the function of each nerve

- the order of each nerve

The cranial nerves are not simply in a particular order; they are numbered. This immediately suggests the appropriate mnemonic: the pegword mnemonic. And the need to remember some rather strange names, and associate this information with function, suggests another useful mnemonic: the keyword mnemonic. I return to this example in the discussions of those mnemonics.

Before getting to the guidelines for creating good first-letter mnemonics, there's one more aspect we need to consider — *when* first-letter mnemonics are useful.

Glossary

medical — de medicina

reiterating — reiterando, repitiendo

purpose — propósito

precise — preciso, exacto

distinguish — distinguir

similar — similares, parecidos

traditional — tradicional, clásico

advantages — ventajas

rhythm — ritmo

constraint — restricción

numbered — numerado

immediately — inmediatamente

appropriate — apropiado

strange — extraño

discussions — discusiones

guidelines — directrices

cranial nerves: olfactory, optic, oculomotor, trochlear, trigeminal, abducens, facial, auditory, glossopharyngeal, vagus, accessory, and **hypoglossal** — nervios craneales: olfatorio, óptico, motor ocular comun, troclear o patético, trigémino, abducens o ocular externo, facial, auditivo, glosofaríngeo, neumogástrico, accesorio o espinal, e hipogloso mayor

When first-letter mnemonics are a good strategy to use, and when they're not

As I implied at the beginning, first-letter mnemonics are effective because initial letters make good retrieval cues. But there's another critical point that's less obvious.

Several studies have found that first-letter mnemonics are of no particular benefit in helping remember, but in all these cases, students were asked to learn unrelated words[1]. However, one study found there *was* a benefit when the *order* of items became important[2], and further investigation confirmed that while the mnemonic isn't useful for learning new sets of unrelated items, it does help when the items themselves don't have to be learned, but the order of them does[3].

The same study also confirmed that, by an overwhelming margin, the chief type of error made by those using a first-letter mnemonic is that of reversal — that is, confusing an item with another item with the same initial.

So, it seems that first-letter mnemonics are of no particular benefit when order isn't important, although if the items in a list lend themselves to a memorable acronym (such as HOMES), then you should certainly take advantage of that fact. Such opportunities will, however, be rare.

Acrostics are much more widely applicable, but generally less memorable than acronyms.

So when are first-letter mnemonics a good strategy to use? The best time is when you have a relatively short list of items, with very familiar items that all begin with different initials. Items should be related — related items provide a much more limited set of possibilities for your initials, and are thus better retrieval cues.

The results of one study have also suggested that first-letter mnemonics may be more effective for females than males, for whom strategies involving visualization may be superior[4]. I imagine this reflects a broader preference of visual over verbal strategies, rather than a particular indictment of first-letter mnemonics however.

Because of their power as retrieval cues, first-letter mnemonics are also particularly recommended for students who suffer from exam anxiety, with consequent memory blocks (having your mind go blank when you look at the exam questions).

Teachers and writers might also like to note a study that found that giving the initial letters of the main points of a concrete (but not an abstract) passage, either before or after the students read the passage, helped them retrieve the main points. Like other strategies that aid memory for particular items, it was however at the expense of remembering other details[5].

First-letter mnemonics are

- a cueing strategy, not a learning strategy
- useful when remembering the order is critical
- useful when you want to prevent a memory block

Principles for creating effective acrostics

Unfamiliar items need more cues. If the items are well-known, and the acrostic is only needed as a reminder or to provide order information, choice of words is only constrained by initial letter. However, if the items are not well-known, the words must also provide cues to the items.

Choose familiar words. Where possible, the words chosen should be familiar words (which are more easily recalled).

Make it meaningful. As much as possible, the acrostic itself should make coherent sense (a meaningful sentence is remembered more easily).

Cue the order. If the acrostic is providing a particular type of order, where more than one type is possible, then it needs to also contain cues to what kind of order is involved.

Keep it simple (don't force your mnemonic to carry too much information). If the acrostic is required to bear information about order, kind of order, and item content, it is usually better to create more than one mnemonic.

Glossary

implied — impliqué

unrelated — no relacionado

overwhelming — abrumador

reversal — inversión

applicable — aplicable, capaz de aplicar

related — relacionado

preference — preferencia

indictment — acusación

exam anxiety — ansiedad de examen

consequent — consiguientes, resultantes

mind go blank — mente en blanco

concrete — concreto, se refiere a cosas físicas

abstract — abstracto, se refiere a conceptos no físicos, como la justicia o el amor

passage — pasaje de texto

at the expense of — a expensas de

Review 2.1

1. Which of these are acronyms?

   a) Every Good Boy Deserves Fruit

   b) ROY G. BIV

   c) NATO

   d) Simon Makes Herons Eat Olives

   e) HOMES

2. Which of these are acrostics?

   a) FACE

   b) My Very Eager Mother Just Served Us Nine Pizzas

   c) HOMES

   d) MRS GREN

   e) Father Charles Goes Down And Ends Battle

3. Acronyms are easier to create than acrostics   T / F

4. Acronyms and acrostics are a good strategy

   a) when you don't know the information

   b) when order doesn't matter

   c) when you just need a reminder

   d) when the specific order needs to be remembered

e) when the words start with only a few different initials

   f) when the words in the sequence are similar

5. To make good acrostics
   a) use bizarre words
   b) make interesting, nonsensical sentences
   c) use ordinary words
   d) make meaningful sentences
   e) don't make it carry too much information

Rhythm & rhyme

Some familiar mnemonic jingles

> Thirty days hath September,
> April, June and November.
> All the rest have 31,
> Excepting February alone,
> And that has 28 days clear,
> And 29 in each leap year.

Many of you will know this jingle, or a variant. I learned this in childhood, and to this day, when I want to know the number of days in the month (and it's something that comes up surprisingly often), I run through the first two lines of this jingle.

The first two lines are all that are really needed in most circumstances, once you're familiar with the verse. After all this time, I don't need the mnemonic to tell me how many days February has, but I do still like to make sure I have the vexed 30-31 question right!

Rhythm and rhyme are powerful aids to memory, but the reason this jingle comes so easily to mind is not because of those two factors alone— it's because of the frequent repetition I've given it. This is something that needs to be borne in mind with all mnemonics— mnemonics make remembering easier, but they don't obliterate the need for repetition. They simply reduce it. Indeed, you could say one measure of how effective a mnemonic is, is the degree to which it reduces the need for repetition.

Here's another little jingle that will be familiar to many:

> In fourteen hundred and ninety-two,
>
> Columbus sailed the ocean blue

Isn't it remarkable how much more memorable the addition of a simple rhyme and a meter makes this? Compare it to: In fourteen ninety-two, Columbus crossed the Atlantic.

Here's one that might be more familiar to British and Commonwealth citizens, commemorating Guy Fawkes Day:

> Remember remember,
>
> The 5th of November,
>
> Gunpowder, treason and plot.

It's all about rhythm and rhyme.

Rhyme is an effective cue to remembering for similar reasons as first-letter mnemonics: because of the way words are filed in memory, and because it provides constraints on the possibilities. Think of songs and poems that you can easily finish the line for, simply because the word seems obvious — because of the constraints of context and rhyme.

That tells us something. It tells us that an important variable in deciding whether a rhyme is effective is the extent to which it is *predictable*. Predictable rhymes, although it may seem counter-intuitive, are generally more effective — hence the banality of most popular and long-lasting mnemonic verses, like those I have just quoted.

Rhythm is a little more complex. Rhythm takes us to music, and perhaps we should start by considering why music helps us to remember. Or at least, *how* it helps us remember.

Glossary

jingle — rítmico sonsonete

variant — variante, variación

verse — poesía, versos

vexed — espinoso

frequent — frecuente

borne in mind — tener en cuenta

obliterate — obliterar, borrar

reduce — reducen

measure — medida

degree — grado

meter — metro

commemorating — conmemorando

filed — archivado

predictable — previsible

banality — banalidad

long-lasting — duradero

complex — complejo, complicado

Singing to remember

Research has convincingly demonstrated that words are more easily recalled when they are learned as a song rather than speech, but *only* in particular conditions. The important thing is the melody.

The reason melody can be useful is, again, because it provides cues to recall — by virtue of the constraints they place on the possibilities, particularly in terms of line and syllable length. So if melody is to be useful, it is crucial that it be simple and predictable (like rhythm).

Simplicity is not a rigid measure of course. What's simple to one person, particularly a musically trained one, may not be simple to another. Simplicity also varies with familiarity — the more often we hear a melody, the simpler it'll become (within limits!).

An important aspect of simplicity is that the text and the music should be closely integrated — specifically, the number of notes in the melody should match the number of syllables in the lyrics.

Simplicity and familiarity are why, if you want to improve memorability by attaching the material you want to remember to a tune, you are best advised to choose a familiar 'nursery' song.

Simplicity also impacts on predictability. Predictability is important for the obvious reason that melody helps us remember text to the extent that it provides cues, but also because we remember expected information better$_1$.

As far as the lyrics themselves are concerned, context is also an important factor. We can predict the words in a song or poem not simply because of the constraints of rhyme and rhythm, but also because the context sets its own restraints. We know that in a love song, *heart* might rhyme with *part* (e.g., *we'll never part*), but is never going to rhyme with *fart* (unless it's a parody!).

This tells us that related text — text that is coherent and meaningful, that belongs together — is going to be much more effective than unrelated text (such as a list of words).

That doesn't mean that unrelated text won't be helped by attaching it to a tune. One study[2] found that, although hearing a list of words sung didn't help people learn the words any better than hearing them spoken, nevertheless, those who heard them sung took less time to relearn the list a week later. However, there are more effective ways of remembering unrelated items — attaching it to music is not going to help much.

There are other reasons why music helps of course — reasons that have to do with motivation and emotion. Music engages us, and singing is fun. We're more likely to repeat a song much more often than a spoken passage. This simple fact — that the number of repetitions is high — accounts for a great deal of our memory for songs.

Attaching words to a melody shouldn't be taken as a magic bullet for remembering. But the fact that making something into a song does make it considerably more pleasurable shouldn't be ignored either.

But if you do try to turn information that you've found hard to remember into a song, remember the rules:

Rules for mnemonic songs

- Simple, well-known tune.
- Words that match the melody, note by note.
- Words framed in predictable, meaningful sentences and phrases.
- Lines that follow a predictable pattern and rhyme where possible.

There are a number of songs that have been written to help with learning science — for example, Flanders & Swann's song describing the First and Second Laws of Thermodynamics, Tom Lehrer's song of the Periodic Table, as well as many more modern songs (did you know there was a Science Songwriters' Association?!) — as well as other subjects (such as Shakespearean prose). I have put links to a number of these on my website at www.mempowered/books/mnemonics-study/resources. You might find it useful to consider such songs in terms of these rules.

A practical example of a teaching song points to some other interesting issues. A study$_3$ was done involving a multimedia instructional module, using a computer animated sequence in conjunction with a song about cellular physiology. The program was presented to 5th and 6th grade students by two different teachers. One of these teachers was happy to sing along; the other refused. There was a clear difference in the effectiveness of this program, depending on whether the teacher modeled the singing behavior or not.

There was also a clear gender difference, with girls being much happier with, and benefiting more from, the use of song. Partly this may be due to gender differences in music processing, but I suspect the main reason for this difference is cultural — the boys thought singing was "uncool" (a belief not helped, of course, by the male teacher refusing to sing himself).

Glossary

research — investigación

convincingly — convincentemente

demonstrated — demostrado
speech — habla
melody — melodía
by virtue of — en virtud de
syllable — sílaba
it is crucial that — es crucial que
simplicity — sencillez
simple — sencillo
integrated — integrado
specifically — específicamente, en particular
lyrics — letra
attaching — adjuntando
tune — melodía
advised — aconsejado
nursery song — canción infantil
impacts — impacta
expected — esperado
parody — parodia
motivation — motivación
emotion — emoción
engages — se compromete
accounts — cuenta
magic bullet — bala mágica
pleasurable — agradable
ignored — ignorado

instructional module — módulo de instrucción

computer animated — animado por computadora

conjunction — en conjunto, juntos

cellular physiology — fisiología celular

modeled — modelado

gender — género

uncool — no genial

thermodynamics — termodinámica

periodic table — tabla periódica de los elementos

associated — asociado

lesson plans — planes de lecciones

prose — prosa

Spoken rhythm

Let's return to rhythm on its own — spoken rhythm.

As with melody, research has had inconsistent results in determining its usefulness as a memory aid, and the reasons are probably the same.

It seems likely that rhythm is helpful to the extent that it:

- creates expectancies,
- sets constraints, and
- makes repetition more pleasurable.

It may also be that spoken rhythm is more likely to be effective as a mnemonic aid when it has a strong musical beat. This would suggest that you will probably enhance memorability if you provide a back beat, most easily by clapping along with your words — but bear in mind that while synchronized physical movement can aid memory, it must be simple enough not to distract from the material you want learned.

It may also be that some rhythms are more effective aids than others. This is suggested by a study[4] that found that reciting the Iliad got the heart beating in time with the breath, which may improve gas exchange in the lungs as well as the body's sensitivity and responsiveness to blood pressure changes (remember that blood flow, through its effect on oxygen flow, is critical for brain functioning). The researchers believe it is the hexametric pace of the Homeric verse that is critical for achieving this effect. They also suggested that such recitation produces a "feel-good effect".

There's also been a study[5] finding that vocalizing the Ave Maria in Latin or a yoga mantra slowed breathing and altered blood flow in the brain.

Dactylic hexameter (dum-diddy, dum-diddy, dum-diddy, dum-diddy, dum-diddy, dumdum), which is the rhythm of classical epics, is unfortunately not one of the more common rhythms in English verse, but its musical counterpart — 6/8 time — is quite common.

However, it should be borne in mind that the most successful (widely used; long-lasting) mnemonic verses are all very short ones. Even the "30 days hath November" verse is one that, personally, I only tend to use the first two lines of, and I confess I had to check my memory of the final three lines on the Web.

The familiar aid to English spelling:

> I before e, except after c

is one that rolls off my tongue with ease, but I didn't know (or had long forgotten) the rest:

> I before E, except after C
>
> And when saying "A" as in Neighbor or Weigh
>
> And weird is weird.

(I also came across this variant, which demonstrates how accurate this "rule" is! — but a useful rule-of-thumb nevertheless:

> I before E, except after C,
>
> with the exceptions of Neither Financier Conceived Either Species of Weird Leisure.)

My point is that, although rhyme and rhythm are useful aids to memory, they are best restricted to very brief jingles. Although long poems have been constructed as mnemonic aids (for example, for learning the English kings and queens, or the American presidents), you do need to put a lot of effort into memorizing them, and there are better strategies for mastering such long lists of facts (which we will get to).

Best Practices for Spoken Rhythm

- Short jingles
- Strong beat
- Simple and predictable
- Enjoyable

Glossary

determining — determinando
usefulness — utilidad
enhance — mejorar
clapping — aplaudiendo
synchronized — sincronizado
physical — físico
movement — movimiento
distract — distraer
reciting — recitando
exchange — intercambio
lungs — pulmones
sensitivity — sensibilidad
responsiveness — reactividad
blood pressure — presión sanguínea
oxygen — oxígeno
hexametric — hexamétrico, tiene una línea de versos de seis pies métricos
recitation — recitación
feel-good — sentirse bien, agradable
vocalizing — vocalizante, diciendo en voz alta
dactylic hexameter — hexámetro dactilico, donde cada pie tiene una sílaba larga y dos cortas
classical epics — epopeyas clásicas

counterpart — contraparte
confess — confieso, admito
forgotten — olvidado
rule-of-thumb — regla práctica, heurístico

Review 3.1

1. Rhymes are easy to remember when they
   a) are predictable
   b) use unusual words
   c) are constrained by a simple rhythm
   d) are repeated a lot

2. Songs are easy to remember when they
   a) are musically complexity
   b) have a simple melody
   c) don't make sense
   d) tie predictable words to a predictable tune
   e) are fun to sing

3. Rhythm helps us remember because it
   a) makes repetition more fun
   b) makes the words more predictable
   c) makes us want to clap or tap our feet
   d) helps regulate our breathing

Part II

Keyword Strategies

The idea of a "keyword" is central to all the more complex mnemonic strategies. At its simplest, the idea reflects the fact that we remember new information by attaching it to information we already know well. This is why experts find it so much easier to learn new information (in their area of expertise) than novices — because they have a mass of related information that is already interwoven with many connections, and to which the new information can readily be connected.

However, not all information can be *meaningfully* connected to your existing information, and this is where the keyword comes in. The keyword is an intermediary that provides artificial meaning.

For example, to remember the name of the famous psychologist Alfred Binet, you could tie the name Binet to *bonnet* (the keyword) and imagine Binet in a bonnet. To remember that *aronga* means *direction* in Maori, you could give the unfamiliar word *aronga* the familiar word *wrong* as a keyword, and tie that to the meaning with the phrase *wrong direction*.

The use of keywords puts these strategies firmly into the category of **transformational elaborative strategies**. Elaboration — adding to and extending the information to be remembered— is one of the fundamental memory strategies, common to both transformational and non-transformational strategies. The term transformational, however, points to (and encapsulates the essence of) mnemonic strategies. This is what mnemonics are all about: transforming information into a form that makes it more memorable.

What you have to bear in mind, however, is that this transformation doesn't in any way add meaning. It will not help you understand the material. Its sole purpose is to add a memorable (but meaningless) connection.

For that reason transformational strategies should only be used when there aren't any meaningful connections (of sufficient memory power) that can be used. That still covers a lot of material!

The keyword method

The keyword mnemonic (first formalized by Atkinson in 1975) is the most studied mnemonic technique, and contains within it the most potential for flexible use in a wide range of learning situations.

The essence of this technique lies in the choosing of an intermediary word that binds what you need to remember to something you already know well.

For example, to remember that the Spanish word *carta* means letter (the sort you post), you select an English word that sounds as close to *carta* as you can get, and you make a mental picture that links that word to the English meaning — thus, a letter in a cart.

Or perhaps, to extend the technique a little further, you want to remember that Canberra is the capital of Australia. *Beer can* is an obvious phrase for Canberra (particularly in light of the Australians' notorious enjoyment of beer!), and you could connect it to Australia by substituting a familiar icon such as a kangaroo or a koala bear. Thus, your image for remembering this fact could be a kangaroo swigging back a can of beer.

We'll discuss the extension of this technique to fact learning in the next chapter. Let's first come to grips with the 'pure' technique, designed for learning new words — most especially for learning a new language, but equally applicable to learning new vocabulary (and every discipline has its own vocabulary).

Glossary

reflects — refleja

novices — novatos

interwoven — entretejido

intermediary — intermediario, actúa como un enlace entre dos cosas

psychologist — psicólogo

encapsulates — encapsula

essence — esencia

formalized — formalizado, dada una forma definida

binds — enlaza

notorious — notorio

swigging — tragando

discipline — disciplina, sujeto

Some examples to practice

Here are some Russian words (transliterated from the Cyrillic alphabet) taken from Atkinson & Raugh's 1975 study. Try to think of English words that sound similar:

gorá	
durák	
ósen	
chelovék	
krovát	
dvor	
kusók	
rot	
naród	
úzhin	

I cruelly asked you to think of similar-sounding words before giving any details about the choosing of keywords, but I wanted you to have the experience of thinking of acoustically similar words before limiting you with that knowledge. On the next page you can see the ones Atkinson & Raugh used.

Note that the important thing is *acoustic* similarity (Atkinson called it the acoustic link). It's all about *sounding* alike, not what the actual letters are.

gorá	garage
chelovék	chilly back
durák	two rocks
ósen	ocean
chelovék	chilly back
krovát	cravat
dvor	divorce
kusók	blue sock
rot	rut
naród	narrow road
úzhin	engine

Nor is it necessary to try and echo all of the word, although these words do. Obviously it is better if it does, but that is not always going to be possible. It is acceptable practice in those cases to echo only part of the word. For example, for *górod* (city), Atkinson & Raugh chose *go* as the keyword.

The next step is to create an imagery link with the meaning. Where it is difficult to create an image, you can make up a sentence or phrase that links the two. Here (on the next page) are the meanings.

gorá	garage	**mountain**
durák	two rocks	**fool**
ósen	ocean	**autumn/fall**
chelovék	chilly back	**person**
krovát	cravat	**bed**
dvor	divorce	**yard, court**
kusók	blue sock	**piece**
rot	rut	**mouth**
naród	narrow road	**people**
úzhin	engine	**supper**

As a general rule, it is the thinking of the keyword that people find hard, not the creation of an image joining the two. But here are some suggestions so you get the idea:

A garage perched on the summit of a mountain.

A jester trapped between two rocks.

A torrent of autumn leaves falling on the wide ocean.

A person shivering as a snowball splashes over his bare back.

A cravat laid out ready for wear on a bed.

In a yard, a judge decrees, and a man and woman separate.

Someone is cutting a large blue sock into pieces.

Deep ruts in a once-muddy road; one curves itself into a smile and then speaks, like a mouth.

A mass of people walking along a road that narrows, squeezing them tightly together.

Supper laid out on an exposed car engine.

If you have diligently tried to form these, or your own, images, you will have noticed that it is not entirely fair to call this an imagery link, or to make a distinction between an image and a sentence. In truth, both words and pictures are important. You need to make sure the crucial words — the keyword and the meaning — are attached to the images. When you picture the people walking along a road that narrows, you need to be thinking *narrow road, people*; when you see the snowball splashing on the person's bare back, you need to be thinking *chilly back, person.* It is after all the words that are critical — the images are there only to trigger the words.

So, you create a picture, and you think clearly about the words attached to the elements in the picture, and every time you recall the picture, you make sure you recall the words attached to the elements.

Another thing I hope you noticed when creating these images, is that the pictures I described were very bare-bones. When you turned my sentence into an image, there will have been details I didn't mention, details that are individual to you, to the way you think and the experiences you have had. That's good.

While you shouldn't get bogged down in crafting very detailed images — indeed, you want to keep the images as simple as possible, so that the crucial elements are not hidden — you do want some details that make the images more memorable for you. How well these images work does depend on you putting in the effort to making them as clear and vivid as you can.

That does *not* mean that you have to be a brilliant visualizer to use this technique effectively!

As we have seen, the technique involves both words and images, which means that those who are highly visual can put more weight on images and those who are more verbal can put more weight on the words. But even those who do not think of themselves as particularly visual do benefit from images, so do make the effort! (I assure you that you will get better with practice).

Let's try another example. Here's a short list of Spanish words for the classroom. Before looking any further, try and come up with English words that sound similar.

el lápiz		pencil
la papelera		wastepaper bin
las tijeras		scissors
la regla		ruler
las cuentas		sums
el pupitre		desk
el techo		ceiling
la pared		wall
el suelo		floor
el pincel		paintbrush

Here are my suggestions (they're not necessarily any better than yours!):

el lápiz	**lapel**	pencil
la papelera	**puppet**	wastepaper bin
las tijeras	**tiger**	scissors
la regla	**regatta**	ruler
las cuentas	**queen**	sums
el pupitre	**pupil**	desk
el techo	**teacher**	ceiling
la pared	**parcel**	wall
el suelo	**swelling**	floor
el pincel	**pencil**	paintbrush

Now we make an image joining the keyword and the meaning:

A big pencil sticking out of a jacket's lapel (where a flower or pin might be).

A puppet hanging over the edge of the wastepaper bin.

Baby tigers running around the classroom waving scissors.

A fleet of yachts on the water, giant rulers in place of their masts.

A queen, complete with crown, frowning over a slate with 2 + 4 written on it.

A pupil, very neatly dressed in an old-fashioned school uniform, sitting at a desk.

The teacher has floated up to the ceiling, and is stuck flat to it.

A big squashy parcel on the wall.

A giant swelling raising the floor.

A pencil with a paintbrush end instead of an eraser end.

Glossary

acoustically — acústicamente
chilly — frío
cravat — corbata
divorce — divorcio
sock — calcetín
rut — surco
narrow — estrecho
engine — motor
sounding — suenan similares
echo — eco
acceptable — aceptable
meaning — sentido
mountain — montaña
fool — tonto
autumn — otoño
bed — cama
yard — patio
supper — cena

creation — creación
perched — posado
summit — cumbre
jester — bufón, payaso
trapped — atrapado
torrent — torrente, inundación
leaves — hojas
shivering — tiritando
snowball — bola de nieve
splashes — salpica
judge decrees — el juez decreta
muddy — fangoso
smile — sonrisa
squeezing — exprimido
exposed — expuesto
diligently — diligentemente
distinction — distinción
bogged down — empantanado
crafting — elaboración
brilliant — brillante, excelente
lapel — solapa
puppet — marioneta
parcel — paquete
swelling — hinchazón
waving — agitando

fleet — flota

slate — pizarra

old-fashioned — anticuado

school uniform — uniforme escolar

floated — flotó

stuck — pegado

flat — plano

squashy — suave

giant — gigante

eraser — goma de borrar

Creating good keywords

There are several things to note about the words I've chosen. First of all, they are all nouns. Moreover, they are concrete, not abstract, nouns. That is, they are things that you can visualize. *Regal*, for example, would be a great keyword for *regla*, except that you can't picture *regal* very well (you could of course picture a king or queen, or someone with great deportment, but would these images necessarily bring to mind the word *regal*? That's the deciding issue.)

Of course, it is not always possible to find an appropriate concrete noun — that's why sometimes you have to go with a sentence or phrase instead of a picture — but that's what you should be trying for first.

They are also all words that are familiar to me. That's where the individual comes in — what is a good word for me is not necessarily a good word for you. You may, for example, find *regalia* a better word than *regatta*; if you're familiar with *lapis* (lazuli) as a gem or pigment, that would be a better word than *lapel*. If you're a birdwatcher, *swallow* would probably be a better word than *swelling*.

Note too, that I chose *queen* for *cuentas*. This goes back to what I said about acoustic similarity: although we don't usually think of *c* and *q* having the same sound, the sound *cu* in Spanish is the same as the English sound *qu*.

Effective keywords need to not only sound like the target word and be able to form memorable links with the meaning, they also need to be different from each other — you don't want to get confused between too-similar keywords. Where possible, they should also be concrete and familiar. (As I've said, you can have a concrete word that symbolizes an abstraction, such as an image of a blindfolded woman holding a pair of scales for *justice*. The crucial thing is that the symbol recalls, for you, the abstract noun.)

Perhaps most importantly of all — more important, research suggests, than distinctiveness, vividness, concreteness — is relational and semantic information. This is why the emphasis now is on making *interactive* images or sentences, in which the keyword and definition interact in some way. It is not enough for the two images, the keyword and the meaning, to be in the same image; they must interact.

Thus we don't have a garage *and* a mountain, we have a garage *teetering on* a mountain. The pupil *sits at* the desk; the puppet *hangs out of* the wastepaper bin. The better (the more active; the more meaningful) the interactive connection, the more effective it will be.

The advantage of a semantic connection may be seen in the following example, taken from an experimental study[1]. Students in a free control condition (those told to use their own methods to remember) almost all used a keyword-type technique to learn some items. But unlike those in the keyword group, the keywords chosen by these subjects typically had some semantic connection as well.

Thus, for the Spanish word *pestana*, meaning *eyelash*, several people used the phrase *paste on* as a link, reflecting an existing association (pasting on false eyelashes). The keyword supplied to the keyword group, on the other hand, was *pest*, which has no obvious connection to eyelash. (It is also worth noting that verbal links were more commonly used by control subjects, rather than mental images.)

It seems likely that keywords that are semantically as well as acoustically related to the word to be learned will be remembered longer and more easily.

Relatedly, two studies[2] that directly compared learning by context (a very popular strategy among teachers) and the keyword strategy found that the best method was one that combined both — that is, where students were given not only the keyword, but also example sentences, showing the word in context.

In other words, use of the keyword mnemonic should not blind you to the value of seeking and attaching meaning. Creating meaningful links should always take precedence over arbitrary links, however vivid and distinctive.

Note too, that although it is usually recommended that you should try to create bizarre images, the research evidence for this is mixed[3]. Having bizarre images seems to help remembering immediately after learning only when there is a mix of bizarre and less unusual images, and may not particularly help over the long term at all.

It seems plausible that one reason for the conflicting

experimental results is the fuzzy question of what exactly constitutes 'bizarreness'. It may be that some forms of peculiarity help memory, while others don't (and may even hinder it).

It does seem clear, however, that most people find it harder to come up with bizarre images. Accordingly, I would recommend that you should only use bizarre images when they come easily to mind.

These, then, are the dimensions along which keywords can be evaluated for effectiveness:

- meaningfulness
- acoustic similarity
- imageability
- distinctiveness
- familiarity

The fact that I call these dimensions is a clue that keywords can, and will, vary considerably along these dimensions! In other words, you try the best you can, but sometimes you will have to accept keywords that measure poorly on these dimensions. That does not necessarily mean they will be ineffective.

For example, on the next page you can see some keywords Atkinson & Raugh used in their study that were startlingly effective. For all of these the difference in recall between those learning in the keyword condition and the control condition was greater that 40% — for example, the probability that those in the keyword condition would remember that *dévushka* meant *girl* was 100% but only 50% for those in the control condition; the probability that those in the keyword condition would remember that *vózdux* meant *air* was 77% compared to 35% for those in the control condition.

dévushka	dear vooshka	girl
vnimánie	pneumonia	attention
taréika	daddy elk	plate
karandásh	car run dash	pencil
edá	ya die	food
vózdux	fuzz duke	air

It cannot be said that these are good examples of keywords according to most of the dimensions I have described, and yet they worked very well. My own feeling is that these are clear examples of the importance, and effectiveness, of the verbal part of the imagery link (Atkinson & Raugh also called this the mnemonic link, and this is perhaps a better term in some ways). Nevertheless, the reason why these keywords were so effective is mysterious, and emphasizes that the guidelines are only that. They are not rules. Ultimately, a good keyword is one that works for you.

Having said that, a study[4] that used "good" keywords and "poor" ones showed not only that the quality of the keywords makes a significant difference, but that judges could accurately assess the memorability of keywords. So, as a general rule (inexplicable cases aside), you can probably usually tell whether a particular word will make a good keyword.

Have a look at at all the words for which the difference in recall between those learning in the keyword condition and the control condition was greater than 40% (opposite page). The first word in this list had a difference of 62% (the probability that those in the keyword condition would remember that *dvor* meant *yard* was 81% but only 19% for those in the control condition), while the last word had a difference of 42% — 65% vs 23%.

Russian	Keyword	Meaning
dvor	divorce	yard
nachálo	not shallow	beginning
kusók	blue sock	piece
vnimánie	pneumonia	attention
dévushka	dear vooshka	girl
naród	narrow road	people
tolpá	tell pa	crown
gorá	garage	mountain
gálstuk	gallstone	necktie
plóshchad'	postage	square
uslóvie	Yugoslavia	condition
durák	two rocks	fool
ósen'	ocean	autumn/fall
taréika	daddy elk	plate
karandásh	car run dash	pencil
straná	strawman	country
edá	ya die	food
vózdux	fuzz duke	air
výxod	boyhood	exit
zházhda	judge	thirst
gólod	gullet	hunger

In other words, those using keywords were still three times as likely to remember the word than those in the control condition.

Interestingly, the study found no particular disadvantage to using a keyword phrase — these were recalled about as well as single keywords. Parts of speech are another matter.

In a later study by Atkinson & Raugh[5] in which verbs and adjectives were used as well as nouns, while nouns and verbs were remembered at about the same rate, adjectives were noticeably harder to remember. While this could be related to the adjectives being on average longer words, it seems most likely to be related to the greater difficulty in visualizing them.

Is it all about good keywords? How important is the image? Atkinson & Raugh explored these questions a little further by adding two groups — those who only learned the acoustic link, and those who only learned the imagery link. From this study it appears that both the acoustic link and the imagery link are important, and neither one is more important than the other. Moreover, the two links are independent — performance on one was unrelated to performance on the other.

But don't forget that the mnemonic link doesn't have to be an image. If a verbal link works better for you, that's fine — just remember that the crucial characteristic is that it connects the keyword with its associated word in a way that is active and meaningful.

Choosing a good keyword

A good keyword will sound as much as possible like the to-be-learned word.

A good keyword will easily form an interactive image with the word meaning.

A good keyword will be sufficiently unlike other keywords to not be confused.

A good keyword will be a familiar word, easily recalled.

Nouns and verbs are usually better than adjectives.

Crafting a good image

A good image will be simple.

A good image will tie the keyword with its associated word (usually the target word's meaning) in a way that is active and personally meaningful.

A good image will be clear and vivid.

Glossary

regal — real

deportment — porte

deciding issue — cuestión decisiva

regalia — insignias reales

gem — joya

pigment — pigmento

birdwatcher — observador de aves

swallow — golondrina

blindfolded — con los ojos vendados

scales — equilibrio

justice — justicia

distinctiveness — diferencia, carácter distintivo

vividness — viveza

concreteness — concreción

relational — relacional

semantic — semántico, relacionado al significado

definition — definición

teetering — tambaleante

hangs out of — cuelga de

control condition — condición de control, grupo que se usa como comparación base

typically — típicamente

eyelash — pestaña

paste on — pegar en

seeking — buscando

precedence — precedencia, prioridad

mixed — mezclado

unusual — raro

conflicting — contradictorio

fuzzy — borroso

peculiarity — rareza

hinder — impedir

dimensions — dimensiones, aspectos

mysterious — misterioso

quality — calidad

significant difference — diferencia significativa

inexplicable — no se puede explicar

How effective is the keyword method?

The keyword mnemonic should not be regarded as a universal remedy. In the Atkinson & Raugh study there was considerable variation among the 120 words to be learned, ranging from the 100% recall for *dévushka* to 27% recall for *lápa* (paw) — which was in fact remembered considerably better in the control condition (58%).

Nevertheless, over a third of the words were remembered over 80% of the time in the keyword condition, compared to only *one* item in the control condition (*glaz* for *eye* — a mnemonic link so obvious I am sure most of the control participants used it). Moreover, only seven words were remembered *less* than half the time in the keyword condition, compared to 70 in the control condition!

Overall, the keyword group recalled 72% of the words when they were tested on the day following the three study days (40 words were studied each day), compared to 46% by the control group. When they were (without warning) tested again six weeks later, the keyword group remembered 43% compared to the control group's 28%.

These statistics make a compelling case for the effectiveness of the keyword mnemonic over rote repetition. The precipitous drop in both cases at six weeks is entirely expected, and should be taken as a loud warning that you cannot expect *any* memory strategy to work without sufficient repetition. I discuss the best way of providing that in the final section.

It should also be noted that the control group was instructed to use any strategy they could think of to learn the words. Assuredly some of the participants used a keyword-type mnemonic at least some of the time — for example, in the case of *glaz* (eye), which was remembered 92% of the time by control subjects.

In comparison, in an earlier study[6] using Spanish words, in which control subjects were specifically instructed to use rote repetition, they only scored 28% compared to the keyword group's 88% (this is decidedly higher than the 72% scored in the Russian study, and may reflect good keywords being easier to find in Spanish, a much more closely related language to English than Russian).

One benefit the keyword method seems to grant is to make the imageability of the word-to-be-learned less important. For the control group, words of high imageability were learned better than words of low imageability, as we would expect. However, for those using the keyword method, the imageability of the words was not an issue — again, unsurprisingly once you think about it, because what the method has done is to create imageability where there is little.

But I'm not advocating total reliance on the keyword method. The first strategy in learning a new word should always be to look for a familiar root or cognate.

Spanish, for example, has many hundreds of common words that are very similar to English. It would be pointless to use the keyword mnemonic on such as those. Indeed, it's worth making an effort to look for roots even when they don't hit you in the face — they're not always obvious, but having found them they become more obvious, and are likely to help you remember the word. (Like many others I am pleased to have learned Latin at school, for the help that has given me in learning those languages that, like our own, includes many words that derive from Latin.)

Glossary

remedy — remedio

statistics — estadística

compelling — irresistible

precipitous drop — caída precipitada

instructed — instruido

assuredly — ciertamente

imageability — capacidad de imagen

advocating — abogando, defendiendo

reliance — confianza

root — raíz

cognate — cognado

pointless — inútil

Limitations of the keyword method

The keyword method is primarily a strategy for recognition learning. You see the word *carta*, the keyword *cart* is triggered, and hopefully the image of the letter in the cart is then recalled. The method is not so useful the other way around, for remembering the Spanish for *letter*.

The problem is, of course, that generating the unfamiliar word from the keyword is much harder than remembering the (familiar) keyword from the unfamiliar word — *go* from *górod* is easy; *górod* from *go* is much less so.

That's why the keyword method is only one of what should be several strategies that you need to employ in learning new words.

But being able to recognize words, to remember their meanings, is hugely useful. Doing that enables you to read more fluently; reading as much as you can is how you get the repetition you need to firmly embed the words (this is a much better form of repetition than rote repetition, because it is more interesting and more contextual). The big advantage of the keyword method is that it gets the words into your head faster, enabling you to more easily get the practice you need.

Experimental research, of course, invariably involves very limited numbers of words to be learned. While this is entirely understandable, it does raise the question of the extent to which these findings are applicable to real world learning situations. If you are learning a new language, you are going to have to learn at least 2000 new words. Does the keyword mnemonic hold up in those circumstances?

While the keyword mnemonic has been used in real world situations (intensive language courses), these are not experimental situations, and we must be wary of the conclusions we draw from them. The keyword strategy does take time and effort to implement, and may well have disadvantages if used to excess. As I've discussed, some words lend themselves to other techniques. At least for more experienced students (who will have a number of effective strategies, and are capable of applying them appropriately) the keyword strategy is probably best used selectively.

Glossary

recognition — reconocimiento

triggered — activado

generating — generando

unfamiliar — desconocido

recognize — reconocer

enables — habilita

fluently — fluido

embed — incrustar

contextual — apropiado al contexto habitual

invariably — invariablemente

understandable — comprensible

intensive — intensivo

wary — cauto

conclusions — conclusiones, resultados

implement — poner en práctica

disadvantages — desventajas

selectively — selectivamente

Remembering for the long term

No one can deny the effectiveness of the keyword method for immediate recall; however, whether or not it is better for long-term recall (which is, after all, what most of us are after) has been a much more contentious issue[7]. While many studies have found good remembering a week or two after learning using the keyword mnemonic, others have found that remembering is no better than any other strategy one or two weeks later. Some have found it worse.

It has been suggested that, although the keyword may be a good retrieval cue initially, over time earlier associations regain their strength and make it harder to retrieve the keyword image. This seems very reasonable to me— any keyword is, by its nature, an easily retrieved, familiar word; therefore, it will already have a host of associations. When you're tested immediately after learning the keyword, this new link will of course be fresh in your mind and easily retrieved. But as time goes on, and the advantage of recency is lost, what is there to make the new link stronger than the other, existing, links? Absolutely nothing — unless you strengthen it. How? By repetition.

Notwithstanding Atkinson & Raugh's finding that the acoustic and the imagery links are equally important (and confirming the relative independence of these two links), over time it seems that it is not the keyword itself that is usually forgotten. It's the image.

This implies that the part of the mnemonic that you must particularly work on is the link between keyword and image. For example, the garage *on the mountain*. It's a natural impulse to put one's energies into strengthening *gorá — garage*, but your focus should be on putting that garage on the mountain.

I believe the reason why the keyword strategy has sometimes shown itself less effective when tested for long-term memory is the ease with which words are initially mastered. This encourages students to give up repeating them long before they have been truly fixed in memory.

What you need to remember is that how long something is remembered for is not a matter of the method you used in learning it — what matters is how *well* you learned it.

It's often suggested that a mnemonic you've thought up yourself will be stronger than one that is given to you, but there is no evidence for this in relation to keywords. Indeed, much of the time students do better if they're given the keyword, rather than having to think it up themselves[8].

It seems likely that the inconsistency in results reflects the fact that whether or not it's better if you're given keywords or better to think them up yourself depends not on any specific rule but on how difficult it is to find an appropriate keyword and how skilled you are.

Similarly, whether or not it's better to see an actual visual image rather than simply have that image described to you, probably depends on the individual — I personally am much more verbal than visual, and I do find constructing an image from clipart helps me considerably. However, those who are good visualizers are less likely to need such assistance.

What is clear from the research is that instruction in the technique is vitally important. Indeed there is some evidence that effective use of the keyword mnemonic requires individual instruction (as opposed to group instruction). But I don't think this is the crucial ingredient to effective instruction.

I think that the difference between experiments where the keyword mnemonic has been clearly superior, and those where it has not, comes down to how much direction the students have been given in how to use the technique.

Durable keyword images do require quite a lot of practice to create. It has been suggested that initially people tend to simply focus on creating distinctive images. It may only be with extensive practice that you become able to reliably create images that effectively *integrate* the relational qualities of the bits of information.

What all this suggests is that **successful and continued use of the keyword mnemonic requires more instruction and practice than you might think**.

Glossary

contentious — contencioso

associations — asociaciones

retrieve — recuperar

host of — gran cantidad de

recency — actividad reciente

absolutely — absolutamente

notwithstanding — a pesar de

impulse — impulso

energies — energías

focus — atención

mastered — dominado

fixed — fijo

assistance — asistencia, ayuda

durable — durable, duradero

extensive — extensa

reliably — de manera confiable

Comparing the keyword mnemonic to other strategies

As a general rule, experimental studies into the effectiveness of the keyword mnemonic have compared it to, most often, rote repetition, or, less often, "trying your hardest to remember" (i.e., your own methods). It is not overwhelmingly surprising that the keyword mnemonic should be superior to rote repetition, and comparisons with "free" controls will often show inconsistent (and uninformative) results because some participants will be using the keyword method at least some of the time.

For example, in one experimental study[9], 17 out of the 40 control subjects used the keyword method for at least some items, and many of the keyword subjects didn't always use the keyword method. For the control subjects, the probability of recalling keyword elaborated items was 81% vs 45% for other items, while for the keyword group, the probability of recall for keyword-elaborated items was 80% vs 16% for those items for which they didn't use a keyword mnemonic.

While this certainly emphasizes that the keyword mnemonic is hugely more effective than rote repetition, it doesn't tell us how the keyword method compares with other effective (or presumed to be effective) strategies.

A number of studies[10] have compared the keyword strategy against the context method of learning vocabulary (much loved by teachers; students experience the word to be learned in several different meaningful contexts). Theory suggests that the context method should encourage multiple connections to the target word, and so it's expected to be a highly effective strategy.

However, the studies have found that the keyword method produces better learning than the context method, including when the students had to work out the meaning of the word themselves, from the context.

This was true even when subjects were given a test that would be thought to give an advantage to the context method — namely, subjects being required to produce meaningful sentences with the target words.

But it seems likely that all this depends on the student and her background knowledge. All these strategies — using context, finding roots and synonyms/antonyms, generating meaningful sentences — are elaboration strategies. It would not be surprising if, as with elaborative interrogation, these strategies require a certain level of knowledge and understanding to be used effectively.

The crucial thing about mnemonics is that they are the best strategy when such knowledge and understanding is lacking.

Glossary

uninformative — eso no da información

elaborated — elaborado

multiple — múltiple

target — objetivo

background knowledge — conocimiento de fondo

lacking — carente

Tasks for which the keyword method is useful

As I said before, it is not only for learning another language that the keyword mnemonic is useful. It is also an excellent strategy for learning new technical vocabulary. Remember when I talked about the common medical mnemonic for the cranial nerves (**O**n **O**ld **O**lympia's **T**owering **T**op **A** **F**inn **A**nd **G**erman **V**ault **A**nd **H**op), I mentioned the problem of remembering the names of the nerves: olfactory, optic, oculomotor, trochlear, trigeminal, abducens, facial, auditory, glossopharyngeal, vagus, accessory, and hypoglossal. Several of these are familiar words, but the rest certainly need something to make them more memorable! See if you can come up with good keywords for them.

We will return to this mnemonic when we look at the pegword method.

oculomotor	
trochlear	
trigeminal	
abducens	
glossopharyngeal	
vagus	
hypoglossal	

Using the keyword mnemonic to remember gender

One other aspect of vocabulary learning for many languages is that of gender. The keyword mnemonic has successfully been used to remember the gender of nouns, by incorporating a gender tag into the image[11]. This may be as simple as including a man or a woman (or some particular object, when the language also contains a neutral gender), or you could use some other code — for example, if learning German, you could use the image of a deer for the masculine gender (der), the iconic image of Death with a sickle or a single die (plural dice) for the feminine gender (die).

Non-European languages

The use of the keyword method in learning vocabulary is obvious when the vocabulary is in a related language; its use between unrelated languages that don't use the same script is much less obvious. Intriguingly, some Chinese researchers have tackled this problem for Chinese learning English[12]. Their solution is to devise an intermediary code, by which every letter in the alphabet is linked to a specific Chinese character that has some phonetic or graphic link to the letter. These associations must, of course, be extremely well learned.

Having over-learned these, the student then goes on to learn what the researchers call the basic key-letters method, which is used only with English words 2-3 letters long. In this, both the letters in a 2-letter word, and the last 2 letters in a 3-letter word, are transformed into their Chinese character counterparts. The meaning of these characters is then integrated into an image or sentence incorporating the meaning of the English word. Once this basic method has been mastered, the student can extend the method to longer words. The study, involving a group of junior high school students (all of whom were reported as possibly learning disabled), was remarkably successful in improving their learning of English words.

In the next chapter, we will look at further extensions of the keyword method.

Glossary

sickle — hoz

dice — dados

tackled — abordado

phonetic — fonético, relacionado con el sonido de las letras

graphic — gráfico

over-learned — sobre-aprendido

learning disabled — con una dificultad de aprendizaje

Main points to remember

Keywords provide artificial meaning when words can't be meaningfully connected.

Practice should focus on the mnemonic link (usually an image) rather than the keyword itself.

The keyword method is mainly a recognition strategy. The keyword method is an effective strategy for learning words fast.

Don't let quick fluency fool you into thinking you have learned a specific keyword mnemonic — long-term learning requires repeated review over time.

Review 4.1

1. A keyword is

   a) an anchor word that helps us remember complex information

   b) a word that is meaningfully connected to a new word

   c) a word that sums up the most important point in a paragraph

   d) a familiar word that can be linked to an unrelated new word through its sound or appearance

2. What's the keyword in this mnemonic image?

   a) letter

   b) carta

   c) cart

   d) trolley

carta = letter

3. What's the keyword in this mnemonic image?

   a) fool

   b) rock

   c) cards

   d) two rocks

durák = fool

4. The best mnemonic images are

   a) vivid

   b) interesting

   c) interactive

   d) simple

   e) distinctive

   f) meaningful

5. The best keywords are

   a) meaningful

   b) concrete

   c) abstract

d) novel

e) familiar

f) distinctive

6. The most important attribute of a good keyword is that it be

   a) concrete

   b) distinctive from other keywords

   c) novel

   d) meaningful

   e) acoustically similar to its linked word

7. The keyword method is useful for:

   a) learning lists

   b) learning the order of familiar items

   c) learning foreign vocabulary

   d) learning technical words

   e) memorizing poems or speeches

8. When using the keyword method, you need to pay most attention to memorizing the

   a) keyword

   b) acoustic link

   c) integrated image

   d) word to be learned

Extensions of the keyword method

More than words

The keyword method was originally designed for the learning of foreign languages, and most of its educational use beyond that has involved its use in specialist topics — learning technical words and concepts in science and social studies. However the technique can be extended to any associated pairs.

For example, one study[1] used a modified keyword method to teach 4th and 5th graders the U.S. states and their capitals. Each state and each capital were given keywords (e.g., *marry* for Maryland; *apple* for Annapolis), and these keywords were linked in a captioned picture ("The capital of Maryland is Annapolis. Here is a picture of two apples getting married"). Because they would ultimately need to recall the capital on seeing the state's name, when learning the capitals the students practiced recalling the capital from the keyword, rather than the other way around. They were taught 12 capital-state pairs using this method, then on the following day they were given 13 new pairs and told to learn them any way they wished. When tested two days later, an average of 71.2% of the keyword-learned pairs were correctly recalled, compared to 36.4% of the control pairs — in other words, around twice as many were recalled using the keyword method.

You could also extend this technique to such associated pairs as the authors of particular books. For example, to remember that Luigi Pirandello wrote *Six Characters in Search of an Author*, you could visualize *piranha* tearing apart *six carrots*. To remember Joseph Conrad wrote *Heart of Darkness*, you could visualize a *con*vict eating a *black heart*. To remember Anton Chekhov wrote *Uncle Vanya*, you could visualize your *uncle* with a *van* tattooed on his *cheek*.

Glossary

originally — originalmente

designed — diseñado

educational — educativo

specialist topics — temas especializados

concepts — conceptos

social studies — ciencias sociales

modified — modificado

captioned — subtitulado

ultimately — por último

convict — convicto, presidiario

tattooed — tatuado

Applying the keyword method to text

A few studies[2] have extended the keyword method to text, mainly to short biographical passages. Here's an example from these studies: to remember that a (fictional) person called Charlene McKune was famous for having a counting cat, the name "McKune" is given the keyword *raccoon*, and an interactive image is constructed combining the counting cat with the raccoon (say, a cat counting raccoons jumping over a fence).

These studies have shown that the keyword method can be an effective strategy for short text. But another study[3] shows that some ways of applying the strategy are better than others.

This study, involving 160 university students, was looking at ways of reducing interference between similar bits of information — for example, the causes of the American Revolutionary War and the War of 1812. In 'ordinary' learning, interference is reduced if there's a significant gap (at least a day) between reading the similar passages. It can also be reduced if you can connect the separate passages to existing knowledge (that's why experts are much less prone to interference when acquiring new information). However, if you had that degree of knowledge in the topic, you wouldn't need to be looking at mnemonics! So it's good to know that mnemonics can also help you minimize interference.

The study used biographical passages that contained 11 pieces of concrete, easily visualizable facts, presented in separate sentences. One group of students was taught a single image integrating all 11 keywords for each passage; another was given 11 separate images (there were also two other control groups, one of whom was given passages that were completely dissimilar).

Here's an example from the study (keywords italicized).

Separate mnemonic:

a *RACCOON* waving from an *apartment* doorway

a *RACCOON* riding a *toy train*

a *RACCOON* talking to a *parrot*

a *RACCOON* throwing *newspapers* onto a doorstep

a *RACCOON* strumming a *guitar*

a *RACCOON* saluting a *soldier*

a *RACCOON* driving a *truck*

a *RACCOON* sucking on a piece of *candy*

a *RACCOON* swinging a *tennis* racquet

a *RACCOON* lying in a pile of broken *pottery*

a *RACCOON* hiking through a *desert*

Integrated mnemonic:

a huge *RACCOON* waving from an *apartment* doorway

to a *parrot* riding a *toy train*

and throwing *newspapers* to a *soldier*

who is strumming a *guitar* while driving a *truck*

loaded with *candy* that has run over a *tennis* player

now lying in a pile of broken *pottery* painted with *desert* scenes

The integrated mnemonic produced as little interference as the control group given dissimilar texts, and significantly better memory than the separate mnemonics (an average 86.1% correct recall compared to 65.7% for the separate mnemonics, and 68.9% for the control using the same texts). As can be surmised from these figures, the separate mnemonics produced as much interference as the control group using the same texts.

All this would seem to be clear evidence that a single integrated mnemonic should be used for related information. However it should be noted that a previous study[4] by these researchers found that separate mnemonics were as effective as an integrated mnemonic when students were asked to learn four 5-sentence biographies. It may be that separate mnemonics are adequate when the number of facts is low.

But it may also reflect how the mnemonics were constructed. In the study where no difference was found, the integrated image was built up bit by bit, one sentence at a time. In the later study, where the integrated image was more successful, the whole passage was presented, and the integrated image shown, in its entirety, at the end.

This seems to suggest that it is better to create a single, integrated mnemonics for related information from text, and to create it as a whole once you have gathered all the information.

Here's another, more naturalistic example of applying the keyword method to text. In this study[5], college students applied the strategy to a 1,800-word passage about historical theories of human intelligence.

The text involved two sub-topics: measurement of intelligence, and structure of intelligence. The former included information about five theorists, and the latter seven theorists. The "free-study" group were given a written summary after each paragraph, highlighting each theorist's major contributions. The mnemonic group were given a keyword for each theorist's name and, after each paragraph, a drawing connecting the keyword with major aspects of the theory (a written description of the picture was also supplied).

So, for example, the paragraph on Alfred Binet was followed by a drawing in which a *crowd* (*crowd* was used to signify the measurement subtopic; a *single person* signified the structure subtopic) watched a man in a racing car wearing a *bonnet*. Under the drawing were the words Binet — bonnet, followed by the text: "This race-car driver is competing in a race while wearing a special bonnet to protect his brain, to remind us of the fact that Binet believed higher mental processes existed and should be measured."

Similarly, the drawing for Spearman showed a single man holding a "primary" *spear*, with several specialized ones on the ground, signifying Spearman's theory of general-plus specific abilities.

Students were tested by writing brief essays, followed by a matching task in which they matched the name with various facts.

Those using the mnemonic technique performed significantly better (more than twice as well) at matching the theorist with the major facts, and about the same as the free-study group at matching them with other, incidental facts. They also did better at knowing the temporal order, although that was not something explicitly emphasized. Moreover, it is noteworthy that both groups showed the same level of structural coherence in their essays. Mnemonic techniques have been criticized for "cluttering" the mind with unconnected facts. This finding counters that criticism.

As in the case of the state-capitals, it pays to think about what your retrieval cues are likely to be. You'll remember that in the last chapter I talked about the problem of backward recall — the fact that it is easier to derive the keyword from the target word (*go* from *górod*), than to derive the target word from the keyword (*górod* from *go*). With the capitals, the students practiced recalling the capital from the keyword in order to overcome this problem.

Similarly, when you're constructing a multi-item mnemonic, you should give some thought to what the retrieval cue is likely to be.

In the study just described, for example, it was assumed the theorists' names (Binet, Spearman, etc) would be given. The student would then retrieve the keyword from the name, and the keyword would trigger the associated image with the rest of the information.

However, it is quite likely that in normal circumstances you would have to recall the theorists' names having been given some general cue, such as "famous psychologists who contributed to the measurement of intelligence". In other words, it is the concept of *measurement of intelligence* that will be the cue, and it is the link between this concept and its keyword that you'd need to practice (as well, of course, as the image linked to the keyword).

Main points to remember

The keyword method can be used for any associated pairs, such as capitals of countries or authors of books.

The keyword method can also be used for several items of related information, such as the main points in a text.

If using the keyword method to remember several items of related information, you should aim to create one single integrated mnemonic.

When practicing your mnemonic, think about what your retrieval cue will be.

Glossary

biographical — biográfico
fictional — ficticio
combining — combinando
applying — aplicando
revolutionary — revolucionario
acquiring — adquisidor
minimize — minimizar
parrot — loro, papagayo
strumming — rasgueo
guitar — guitarra
pottery — cerámica
hiking — excursionismo
dissimilar — disímil, diferente
separate — separado
surmised — conjeturado
figures — cifras
biographies — biografías
adequate — adecuado
facts — hechos
entirety — totalidad
gathered — reunido
naturalistic — naturalista
theories — teorías

sub-topics — subtemas

measurement — medición

structure — estructura

former — primero

theorists — teóricos

latter — segundo

highlighting — destacando

contributions — contribuciones

major aspects — aspectos principales

supplied — se proporcionó

signify — significar, indicar

bonnet — capó

incidental — no dirigido

temporal order — orden temporal, organización de eventos en el tiempo

explicitly — explícitamente, declarado abiertamente

noteworthy — digno de mención

structural coherence — coherencia estructural

criticized — criticado

cluttering — abarrotando

unconnected — desconectado

finding — hallazgo

counters — contrarresta

criticism — crítica

derive — derivar

overcome — superar

The face-name mnemonic

You probably haven't thought about this mnemonic as one applicable to study tasks, although you may well be familiar with it. The face-name association method is one of the more commonly used mnemonics, and it is an extension of the keyword method. You begin by selecting a distinctive feature of a face and searching for a word or phrase that is acoustically similar to the person's name. You then create an interactive image linking the feature with the keyword.

This mnemonic is of course mostly used to help people with that most-important everyday task of remembering people's names. However, the method has applications beyond that. Carney & Levin[6] have explored using this mnemonic to help art appreciation students remember which artists painted what paintings.

Applying the face-name mnemonic to art & artists

The application of the face-name technique to paintings (or indeed other artwork) is readily apparent:

1. Select a distinctive and prominent element in the art.
2. Search for an acoustically similar word to the artist's name.
3. Create an interactive image (or sentence) connecting the keyword to the element.

Thus, for the famous painting of "Gilles", a clown in a baggy white suit, by Watteau, you could imagine the clown carrying buckets of water, with water pouring down on him from above (example from Carney & Levin 2000; for a link to an image of this and the Rouault painting mentioned below, go to the book resources on my website). To remember that Monet painted "Water lilies", you could imagine money raining down on, and being gathered up into, a giant waterlily.

This in itself is a very helpful application, but the really intriguing extension is what the researchers did next.

Knowing the artist when you see a particular painting that you have studied is one thing, but even better is seeing a painting you've never studied, and knowing who the artist probably is. That is, being able to recognize an artist's style.

Carney & Levin compared students' performance on remembering the artist of a painting they'd seen before, and correctly attaching an artist to paintings they hadn't seen before, according to four different training instructions. Half the students were told to use their own methods, and half were trained with the mnemonic. However, half of the mnemonic group were told to select something distinctive about the style or theme of the artist, rather than a specific detail in the painting.

For example, for the painting *This Will be the Last Time, Little Father!* by Georges Rouault, the mnemonic-detail group focused on the skeleton, creating an image of the skeleton's ribs being played on with a ruler (keyword for Rouault). The mnemonic-general group, however, focused on the heavy dark lines that are characteristic of Rouault's work, and imagined making heavy dark lines with a ruler dipped in black paint.

Similarly, half the control ("use your own best method") group were told to focus on the general theme or style (e.g., "heavy, dark lines").

The results were very clear. Despite being told to focus on the style, the control-general students were little better than the control-detail students at recognizing new paintings (an average of just over 50% compared to 44%). The mnemonic-general students, however, were clear winners (an average of 73% correct). (The mnemonic-detail students scored just under 52% — that is, no better than the control-detail students.)

Moreover, the mnemonic general group still scored very well on the specific paintings they had studied (90% compared to 99% for the mnemonic-detail group, 73% for the control-detail, and nearly 67% for the control-general).

Applying the face-name mnemonic to animals

A further study[7] by these researchers used this technique to remember the names of animals. For example, the animal *capybara* was given the keyword *cap* and students were told to "Imagine this flat-headed animal with a cap (capybara) pulled down low over its eyes!". Similarly, a long-tailed lizard called a basilisk could be given the keyword *basket*, and an image could be formed in which the lizard is in a small basket, its long tail draped over the side.

Again, the students using the mnemonic performed significantly better than those using their own methods.

This is a useful application for those studying zoology, and indicates further extension to other organisms, for example, plants and even microscopic creatures. Indeed, the researchers suggest the technique could be extended to identifying countries from their outlines, recognizing parts of the body, minerals, and different kinds of dinosaurs.

Extending the mnemonic to taxonomic & attribute information

In an extension of a similar study[8], in which students were taught to distinguish and name different fish species, the students were also given some hierarchical (taxonomic) information about the fish. For example, "The order Scorpaeniformes includes the family Triglidae, which in turn includes the two fish species, Gurnard and Robin."

Those in the mnemonic condition were told to "Imagine that a scorpion has a hold of a tiger. The tiger leaps down toward a guard who is guarding a robin." There was also a picture illustrating this.

The mnemonic group once again significantly out-performed the control group, on matching the fish to the name (89% vs 76%), on identifying the fish (88.5% vs 60%), on filling in blank hierarchical arrangements (80% vs 37%), and on an analogy test of students' ability to infer hierarchical fish-classifications and relationships (77% vs 48%).

In a similar vein, a study involving 8th-grade students[9] demonstrated the effectiveness of this kind of mnemonic for remembering the attributes of minerals. The students were given information about nine different minerals, and their attributes on three dimensions: hard/soft, pale/dark, home use/industrial use. Those in the mnemonic condition were given, in addition to this information, keywords for each mineral name, iconic images for each attribute, and a drawing combining all these. Thus, *hard* was represented by an old man, and *soft* by a baby; *dark* by a mean dark cat, and *pale* by a friendly pale cat; *home use* by a living room, and *industrial use* by a factory. The drawing for the mineral wolframite showed a baby hiding from a mean dark cat riding a wolf in a living room.

Tellingly, however, the group that saw all this in a picture performed not only significantly better than the group using their own method, but also significantly better than the group given the keywords and iconic images but not shown the illustration or given any description of it (they were simply told to "form a picture in your mind" drawing all these things together).

Does this mean that it's not enough to mentally visualize such complex images? Perhaps. But it may be that the students simply lacked sufficient training to use the mnemonic effectively without more support (and remember that these were children). And again, of course, it surely depends on the individual's visualization abilities.

These two examples are similar to the examples of learning from text, in that both involve the construction of multi-item integrated images. It seems likely that creating such complex mnemonics does require considerably more practice than the simpler technique applied to associated pairs.

Main points to remember

The face-name association method can be used to encode visual information such as artwork, and animals (and of course famous people!).

The face-name association method and keyword method can be combined to form complex mnemonics.

Complex mnemonics integrating several bits of information do require a lot of practice.

Don't forget to think about what your retrieval cue will be!

Glossary

face-name association — asociación cara-nombre
extension — extensión
everyday task — tarea diaria
art appreciation — apreciación artística
artwork — obra de arte
apparent — evidente
prominent — prominente, conspicuo
clown — payaso
waterlily — lirio de agua
intriguing — intrigante
artist's style — estilo del artista
correctly — correctamente
theme — tema
skeleton — esqueleto
ribs — costillas
lizard — lagarto
basilisk — basilisco
organisms — organismos
microscopic creatures — criaturas microscópicas
hierarchical — jerárquico
illustrating — ilustrando
out-performed — superado
identifying — identificando

classifications — clasificaciones

relationships — relaciones

in a similar vein — en un sentido similar

attributes — atributos, caracteristicas

represented — representado

tellingly — eficazmente

illustration — ilustración, imagen

support — apoyo

Review 5.1

1. The keyword method can be used for learning

   a) related facts

   b) capitals of countries

   c) any linked pairs

   d) authors & their books

   e) passwords

2. The face-name association method can be used for learning

   a) faces & names

   b) foreign vocabulary

   c) taxonomies

   d) artists & their works

   e) attributes of organisms

3. When using the keyword method to memorize a number of related facts, it's best to

   a) make a lot of separate mnemonic images

   b) make a single, integrated image

   c) create the image bit by bit as you gather the facts

   d) create the image as a whole once you've gathered all the facts

Part III

List Mnemonics

Mnemonics for lists encompass almost all the complex mnemonic strategies after the keyword method (which is however a crucial component of these list-mnemonics).

Although shopping lists are always mentioned in the context of these strategies, and the "classic" list mnemonic (known variously as the method of loci, place or journey method) was developed primarily to help politicians remember their speeches, list mnemonics can also be usefully applied to study situations.

Ordered information, such as the top ten longest rivers, or the steps in a sequence, are obvious candidates. But list mnemonics can also help you remember the main points of a text.

Even though this information is, presumably, meaningfully connected, when you are still building up your knowledge in the area it may well be that you lack sufficiently deep understanding to rely on that for memory. Or it may be that, although you understand it well enough, you need to be able to repeat with reasonable fidelity exactly what this particular text says.

In other words, although you might know all the individual

bits of information well enough, you might need help in ensuring that each one is tied firmly to the next one.

Furthermore, for those of you who have read *The Memory Key* or *Perfect Memory Training*, and remember the emphasis I put on anchors, those key bits of information that serve as reference points for a cluster (a small network of tightly connected information), list mnemonics can help you remember the anchors until your own developing understanding renders them unnecessary.

This is a point I would like to emphasize — building understanding requires time, and in the initial stages there is a lot that needs to be firmly tamped into your brain, before you truly, deeply, understand it. Mnemonics can help you at this stage.

Don't worry about filling your head with meaningless, arbitrary connections, or that these might stand in the way of you developing true understanding. Mnemonics are merely a crutch that soon fades, once you have the information properly stored.

There are two main kinds of textual material for which mnemonic strategies are particularly appropriate:

- text that is readily understandable but which contains a number of details that might be overlooked
- text that is structured, but is not sufficiently well-known or well-organized for the structure to be used as a frame for retrieval.

For your anchors you should select details you suspect you wouldn't otherwise remember, or details that would serve as effective cues for other bits of information. You encode those details by creating a visual image for them, and then integrate the details using the list-learning mnemonic.

> **To use a list-learning strategy for text**
>
> **Select** the anchors.
>
> **Encode** the anchors (keyword mnemonic).
>
> **Cluster** the encoded anchors (list mnemonic).

Let's look at the various list-mnemonics.

Glossary

encompass — abarcan

mentioned — mencionado

candidates — candidatos

main points — puntos principales

fidelity — fidelidad

reference — referencia

anchors — anclas

tamped — apisonado

meaningless — sin sentido

crutch — muleta

overlooked — pasado por alto

frame — marco

suspect — sospecha

The story method

The story method (sometimes called the sentence mnemonic) is the most easily learned list-mnemonic strategy, although it is not as widely known as the other simple methods we've talked about so far.

As its name suggests, the story method involves linking words to be learned in a story. While this is most obviously useful for learning actual lists, it can also be used for remembering the main points of a passage. In such a case, you need to reduce each point to a single word, which hopefully has the power to recall the whole point.

Examples

Remembering word lists

Let's look at an example. First, an easy one — a list:

Vegetable Instrument College Carrot Nail Fence Basin Merchant Scale Goat

This can be transformed into:

A **vegetable** can be a useful **instrument** for a **college** student. A **carrot** can be a **nail** for your **fence** or **basin**. But a **merchant** would **scale** that fence and feed the carrot to a **goat**.

But let's face it, this is not a very probable list of words for you to memorize. The example is taken (with some modification) from a laboratory experiment[1], and the few tests of the story mnemonic that there have been have tended to involve such lists of unrelated words. But learning lists of unrelated words is not something we need to do very often. And generally, if we do have lists of words to learn — say, the names of the elements in the periodic table — they're going to be too technical to lend themselves readily to creating a story.

Even if the words themselves are not particularly technical, the nature of them is not likely to lend itself to a narrative. Let me show you what I mean. Consider the taxonomy of living things:

Kingdom

Phylum

Class

Order

Family

Genus

Species

Here's an attempt at a story:

In the **kingdom**, **phylum** is a matter of **class**, but **order** is a matter for **family**, and **genius** lies in **species**.

The trouble with this is not the re-coding of *genus* to *genius*; the trouble is, it doesn't make a lot of sense. It's a sentence, but not a story — there's no narrative. Humans think in stories. We find them easy to remember because they fit in with how we think. It follows then that the more effective story mnemonics will actually tell a story. To do that, we're going to have to transform our technical words into more common words.

> **King Phillip** went to the **classroom** to **order** the **family genius** to **specifically** name the individual who had stolen the taxi.

The last part of this is of course unnecessary — you could finish it after individual if you wished. But an important thing to remember is that it's not about brevity. It's about memorability. And memorability is not as much affected by amount to remember, as it is by the details of what is being remembered. So meaningfulness is really important. Adding that little detail about stealing the taxi adds meaningfulness (and also underlines what this mnemonic is about: taxonomy).

Here's a longer example. Remember our hard-to-remember cranial nerves? This story was mentioned in a 1973 *Psychology Today* article by the eminent psychologist G.H. Bower[2]:

> At the **oil factory** the **optician** looked for the **occupant** of the **truck**. He was searching because **three gems** had been **abducted** by a man who was hiding his **face** and **ears**. A **glossy photograph** had been taken of him, but it was too **vague** to use. He appeared to be **spineless** and **hypocritical**.

Here it is again with the nerves shown for comparison:

> At the **oil factory** (olfactory) the **optician** (optic) looked for the **occupant** (oculomotor) of the **truck** (trochlear). He was searching because **three gems** (trigeminal) had been **abducted** (abducens) by a man who was hiding his **face** (facial) and **ears** (auditory). A **glossy photograph**

(glossopharyngeal) had been taken of him, but it was too **vague** (vagus) to use. He appeared to be **spineless** (spinal accessory) and **hypocritical** (hypoglossal).

Notice how, with these technical words, they have been transformed into more familiar words — this is what I meant by saying the keyword method is a vital part of all these list-mnemonics.

Remembering text

Let's try something completely different. Here's a selection of points from articles in my local newspaper that I want to remember to tell my partner: council promises support for replacement of the fire-damaged local surf lifesaving clubhouse; council calls for comments about proposal to open a pedestrian mall to buses; daylight saving marks the start of beach restrictions for dogs; Playcentre Federation calls for government support for more parent education classes; robot exhibition coming soon; organizers of eDay (for recycling computer equipment) given award; secondary schools choir to sing at cathedral.

The first thing to do is choose a keyword / phrase to represent each item: burnt house, bus, dog, parent class, robot, computers, cathedral. Now I need to construct a story. A big advantage I have in this case is that the order is not important, which helps a lot. So I can say:

> The **robot** walked out of the **burnt house**, carrying a **broken computer**. Barking, the **dog** herded him onto a **bus**, which took them both to the **cathedral**, where a group of **parents** were having a **class** on computer trauma.

Once again, there are elaborative details which serve no purpose but to make the story more memorable: that the dog is barking; that the class is about computer trauma. Where possible, you should always try and select concrete keywords — ones that are easy to visualize (even though this is a verbal rather than a visual strategy).

Let's try something a little more abstract. Say you're Scott Atran giving a speech on the genesis of suicide terrorism (taken from an article by Scott Atran, published in *Science*, and reproduced in *The Best American Science and Nature Writing 2004*). Here are the main points you want to cover:

Definition (freedom fighters; French Resistance; Nicaraguan Contras; US Congress, act; two official definitions; restriction to suicide terrorism)

History (Zealots; hashashin; French Revolution; 20th century revolutions; kamikaze; Middle East — 1981 Beirut; Hezbollah; Hamas; PIJ; Al-Qaida - Soviet-Afghan War; fundamentalism error)

Difficulties of defending against (many targets, many attackers, low cost, detection difficulty; prevention)

Explaining why (insults; attribution error; Milgram; perceived contexts; interpretation)

Poverty link (crime — property vs violent; education; loss of advantage)

Institutions (unattached young males, normal, personal identity — Palestinians, Bosnians; peer loyalty; emotional manipulation)

Benefits (to individuals, to leaders, to organizations; effect of retaliation)

Prevention strategies (searches; moles; education; community pressure; need for research)

Rather than coming up with concrete keywords, let's try these main terms as they appear:

Toynbee's **definition** of **history fails** because it doesn't **explain why poverty** is dangerous, how **institutions benefit** young men, and how to **prevent** poor young men making history.

Toynbee (a famous historian) was chosen because we always remember things better if they involve an agent — thus ascribing the definition to a person is better than saying "the best definition" or some such phrase.

The trouble with this is that it is not in itself particularly memorable: it doesn't really tell a story, again it's simply a coherent sentence. Worse than that, it's an abstract sentence. Let's try again, substituting our abstract words for more concrete terms:

Taking my **dictionary** in one hand and the **history** book in the other, I **defended** myself fiercely against the **wine** being hurled by the **Franciscan monk** as I passed by the **church**. The building gave him the **advantage** of height and protection, and throwing my books at him only infuriated him. I searched for something else to **protect** myself with.

I tested this out by trying to recall my little story some hours later. I had no problem with the first part; I had to think a little to recall "advantage", and I had to really search for "protect". (I also tried to recall my first, abstract, sentence — this was much harder, and in fact I couldn't get past "explain why poverty".)

Let's try substituting our last two, abstract, words:

Taking my **dictionary** in one hand and the **history** book in the other, I **defended** myself fiercely against the **wine** being hurled by the **Franciscan monk** as I passed by the **church**. But **Benedict Arnold** joined him, throwing **condoms** at me.

If you know anyone called Benedict (or Bennie or Ben), that would be better. Or you could use benefactor, though it would help if you have a specific person in mind, who you immediately associate with benefactor. Condoms, of course, represent prevention / protection.

This of course only represents the main headings, the outline of the speech. To remember the points within each heading, you construct a separate story (or different mnemonic) for each one. Thus, your chain linking freedom fighters; French Resistance; Nicaraguan Contras; US Congress, act; two official definitions; restriction to suicide terrorism, will begin with your dictionary. It's better to construct these separately for several reasons:

- it will be a very lengthy story if you include all the details in one story;
- if you lose your way, your outline and the other stories will be unaffected;
- it's easier to recover if you get derailed by questions.

As a guideline, there is some evidence[3] that there is little benefit from using the mnemonic for nine or more items.

Pros & cons of the story method

There are several points I want to make about the story method. First, it is a strategy that is easily learned.

Reflecting this, a study[4] involving 65 older adults (average age 67 years) found that the story method was initially of greater benefit than the loci method (a more complex mnemonic involving visual imagery, which we will look at in a later chapter).

However, with more practice, the loci method produced superior results. Interestingly, at the end of the training, when they were permitted to choose which of these two methods they wanted to use to learn the last items, more than twice as many chose the loci method. Even more tellingly, those who chose the loci method were the more expert in mnemonics.

These findings point not only to the fact that the story method is easier to learn, but also that it is less powerful. This is not surprising — as a general rule, choosing a mnemonic strategy is a matter of trade-offs. You need to consider:

- how much effort you're prepared to put in
- how difficult the material is to learn
- how much you have to learn.

The more powerful strategies require more effort to master than the simple mnemonics we've been considering so far; it's only worth putting the effort in if you have a large amount of material that needs such effort.

You may be thinking that the reason why the story method is less powerful is that it is a verbal rather than a visual strategy. However, the story method has a direct counterpart among visual imagery techniques (the link method). A comparison of the two methods[5] found that although the imagery method resulted in better recall when people were tested immediately after learning, there was no difference after a week.

Although the story method is not as powerful as some other, more complex, mnemonic strategies, it's still an effective one. A comparison of the story method with three other learning strategies[6] found that although there was no difference in remembering immediately after learning, when tested a week later students who used the story method recalled the most. The benefit was even greater when they were tested two months after the initial learning experience.

The other methods were simple repetition, first-letter mnemonic, and category clustering. It's worth noting that at one week there was no significant difference in recall between those who used the first-letter mnemonic and those who used repetition. Category clustering was, however, of significant benefit. The researchers surmised that the reason category clustering didn't maintain its advantage was that the words to be learned were not particularly appropriate for such a technique, being largely unrelated.

This is a reminder that it is not simply about finding effective strategies — it's about matching effective strategies to the appropriate tasks / material.

An effective story mnemonic

- tells a meaningful story
- uses familiar, preferably imageable, words
- includes elaborative details that help memorability
- is not too long (fewer than 9 items)

Glossary

actual — real, verdadero

lend itself — prestarse

brevity — brevedad

optician — óptico

abducted — secuestrado

glossy photograph — fotografía brillante

hypocritical — hipócrita

replacement — reemplazo

surf lifesaving — de salvamento de vidas de oleaje

clubhouse — casa club

pedestrian mall — centro comercial peatonal

daylight saving — hora de verano

exhibition — exposición

recycling — reciclaje

equipment — equipo

barking — ladrido

genesis — génesis, origen

attribution error — error de atribución, la tendencia a creer que lo que hace la gente refleja lo que son

perceived — percibido

derailed — descarriló

permitted — permitido

trade-offs — compromisos

Review 6.1

1. List mnemonics can be used for remembering

   a) the main points of a speech

   b) technical vocabulary

   c) items in a fixed order

   d) anchor details

   e) passwords

2. The story mnemonic helps us remember because it

   a) creates vivid visual images

   b) takes advantage of our natural ability to remember stories

   c) uses familiar words

   d) links words in a chain

3. "King Phillip went to the classroom to order the family genius to specifically name the individual who had stolen the taxi" is an example of

   a) an acrostic

   b) a first-letter mnemonic

   c) a story mnemonic

   d) the keyword method

4. The best story mnemonics

   a) use concrete words

   b) use familiar words

   c) keep a tight focus on the important words

   d) are short

   e) make sense

The place method

The place method (or journey method, or method of loci, as it is traditionally known) is the classic mnemonic strategy, having its first recorded use 2500 years ago.

First of all, you choose a place you know extremely well. You might use a familiar route, your house, or a particular room in it. The crucial thing is that you can easily call to mind various 'landmarks' (different fixed objects in a room, for example, or different buildings on a route). These landmarks are your anchors. You must train yourself to go around your landmarks in a particular order. With a route of course, that is easy.

To remember a list, you simply imagine each item in turn at these landmarks. For example, a loaf of bread on the couch; a giant apple on the coffee table; the sink full of carrots; a giant banana in the bath, etc.

As with all mnemonics, you have to try it to appreciate that it really does work! The critical thing is to make sure you know your set of landmarks very well. That is, that you can close your eyes and clearly mentally visualize the places you are walking through.

Having done that, you must clearly, vividly, visualize the items you want to remember in those places. How well this strategy works for you does depend on your ability to make mental images.

But if you think you're not very good at this, remember what I said earlier: most people are better than they think, and it is a skill that improves with practice.

Like most mnemonic strategies, you can apply this strategy at different levels of expertise. At a basic level, it is quite easily learned, as long as you have a familiar place that you know in sufficient detail.

That may not be as easy as it sounds. It's quite amazing how unobservant we are! You may have trouble even with your own home, depending on how long you've lived there, how often you've moved, how often you re-arrange or change the furniture, etc.

So the first determinant of whether this is an easy strategy for you is how well you know suitable places. The second determinant is how good your visualization abilities are — this really is a visual strategy, not one that has a verbal counterpart or significant component.

Using the place method

Assuming this is a good strategy for you, the next issue is when to use it. Shopping lists are all very well, but personally I'm just as happy writing things down on the back of an old envelope.

Unlike shopping lists, most lists are more abstract — which means they're harder to visualize, and since this is a purely visual strategy, you must have concrete (i.e., imageable) items. This means you have to transform your abstract items into concrete ones.

But we have seen, in the discussion of the keyword mnemonic, how this can be done. Moreover, not all your lists or texts will be abstract. Literature is usually grounded in visual images. Here, for example, is a brief text from Shakespeare:

> There is a tide in the affairs of men,
>
> Which, taken at the flood, leads on to fortune;
>
> Omitted, all the voyage of their life
>
> Is bound in shallows and in miseries.
>
> On such a full sea are we now afloat,
>
> And we must take the current when it serves,
>
> Or lose our ventures. (Julius Caesar, IV. iii. 217)

Let's dismember it into concrete images:

> tide: a rising tide
>
> affairs of men: male lovers
>
> flood
>
> fortune: a pile of gold
>
> omitted: an oven mitt
>
> voyage: ship
>
> shallows and in miseries: puddles, person weeping
>
> On such a full sea are we now afloat: middle of the ocean, ship sailing serenely
>
> current: currant (the dried fruit)
>
> lose our ventures: vultures flying away.

And now we place these images in their places:

> Open the front door, and there's a tidal wave coming in.
>
> On the inside couch are two males entwined.
>
> Water floods over the coffee table.
>
> The second couch is covered in a pile of gold.
>
> An oven mitt lies on the dining table.
>
> A toy boat sits on the fridge.
>
> There are puddles on the stove.
>
> A child sits on the microwave, crying.
>
> Another toy boat sails in the sink full of water.
>
> A giant currant sits on the TV.
>
> A vulture stands on the closet, wings out, ready to fly away.

Here's another example. Remember Atran's speech on terrorism? This was the story I came up with:

> Taking my **dictionary** in one hand and the **history** book in the other, I **defended** myself fiercely against the **wine** being hurled by the **Franciscan monk** as I passed by the **church**. But **Benedict Arnold** joined him, throwing **condoms** at me.

We can express this same information using the method of loci:

> On the front doorstep there's a dictionary.
>
> On the inside couch there's a history book.
>
> The coffee table is covered by a shield.

There's a bottle of wine on the second couch.

A Franciscan monk is sitting at the dining table.

A model church sits on the fridge.

Benedict Arnold is working at the stove.

The microwave is covered in condoms.

Some advice from antiquity

As mentioned, the place method dates back to antiquity. The principal instruction manual (an anonymous Roman text known as *Ad C. Herennium libri IV*, dated at 86 B.C.E.) gave the following advice regarding its use:

- To help make sure you haven't gone astray in your order, give every 5th and 10th place a distinguishing mark.
- A rarely frequented place is better, because crowds of passing people tend to interfere with the memory.
- The places shouldn't be too similar to each other, otherwise you're likely to get confused.
- The places shouldn't be too large (which makes the images attached to them vague) or too small (the images will be overcrowded).
- The places shouldn't be too brightly lit (which will make the images dazzling) or too poorly lit (which will make the images hard to see).
- The intervals between places shouldn't be too small or too large — they suggest 30 feet.

More modern research[1] suggests the best routes are circular — these seem to reduce serial position effects (position in a list usually influences memory).

These instructions are surprising in the extent to which they specify the visual characteristics of the places that are, after all, to be used in your imagination. But despite this, it is not required that these places actually exist — places that exist only in your imagination are equally acceptable, as long as you can visualize them sufficiently well (these instructions point to how well that needs to be!). Indeed, Dante's Inferno supposedly described the circles of hell in such vivid detail so that they might serve as places that could be used in this way. In modern times, video games might provide such mnemonic places for some.

There is no denying that this method is one that requires far more training and practice than the other methods I've described. You need to not only be well-versed in your particular route (or routes — if you're serious about using this method, you'll find it useful to have several different routes), but also in forming the images that you place on the loci. Moreover (and this is true for all list-mnemonics), if you're going to use the method to help you remember important points in a text, you need to become skilled in identifying words in the text that provide good cues.

Effective loci

- are very well-known
- form a clear, circular route
- are dissimilar enough from each other not to be confused
- are moderately, and reasonably evenly, spaced
- are visualized in clear lighting
- provide sufficient area for the arrangement of items

When to use the place method

In Roman times, this method was popularized as a strategy for helping people remember speeches (remember that classical cultures were predominantly oral; rhetoric was a highly prized skill). This is indeed one of the tasks for which this method is most suitable, because its purely visual nature enables you to hold it in your mind simultaneously as you speak — visuospatial working memory is separate from auditory/verbal working memory. So this method is an excellent choice if you are giving an oral presentation.

It's also a good choice if you want to remember main points for exam essays — particularly if you suffer from exam anxiety, where your brain freezes up. All mnemonics are good for that, but it may be that a visual strategy is particularly effective, given that your exam block is rooted in verbal information.

Interestingly, a test$_2$ of the effectiveness of the method with different kinds of text found that the method was most effective with expository text (compared to descriptive and narrative passages — there was no difference between these two in terms of performance). However, this was only significantly true when the texts were *heard*, rather than read (although performance on written expository text was better, it did not reach statistical significance).

In general, recall was higher, and more affected by passage type and mnemonic method, when the texts were presented orally. For example, the highest average recall was 33.5% (achieved by using subject-generated loci on an expository passage), and the lowest average recall was 14.9% (from using verbal rehearsal on a narrative passage). The highest average recall when the texts were written was 30.6% (achieved by using verbal rehearsal on an expository passage), and the lowest was 17.1% (from using subject-generated loci on a descriptive passage).

The variability of these results emphasizes that it is no simple matter to rule when a specific mnemonic is the best tool for the job. In this particular study, verbal rehearsal was clearly better than the place method for written presentations of both descriptive and expository texts (regardless of whether the loci were generated by the experimenter or the subject). However, when the written text was narrative, both verbal rehearsal and the loci method (where the loci were supplied by the experimenter) achieved the same level (in fact the loci method was slightly superior, but not significantly). And when the text was presented orally (as it would be in a lecture or seminar), the loci method was clearly superior to verbal rehearsal for all types of text — with subject-generated loci always superior to experimenter-generated loci.

It does seem clear, however, that the best time to use the place method is in an oral situation — when you are either listening or speaking.

The place method is particularly useful:

When you are listening to a lecture

When you are giving a speech

For written text, when the text is expository rather than narrative or descriptive

Glossary

landmarks — hitos, puntos de referencia
unobservant — inobservante
grounded — se basa
dismember — desmembrar
tidal wave — oleada, ola gigante
entwined — entrelazado
couch — sofá
puddles — charcos
microwave — microonda
currant — pasa de Corinto
vulture — buitre
closet — armario
antiquity — antigüedad
anonymous — anónimo
astray — extaviado
dazzling — deslumbrante
serial position — posición serial
effects — efectos
well-versed — bien versado
predominantly — predominantemente
rhetoric — retórica
highly prized — muy apreciado
simultaneously — al mismo tiempo

auditory — auditivo

expository text — texto expositivo, texto cuyo propósito es explicar o informar

descriptive — descriptivo, texto que describe a una persona, lugar o cosa

rehearsal — ensayo, repetición

seminar — seminario

Review 7.1

1. The method of loci is a

    a) verbal mnemonic

    b) visual mnemonic

    c) strategy that combines verbal & visual aspects

    d) list mnemonic

    e) transformational elaborative strategy

2. The method of loci uses

    a) keywords

    b) numbers

    c) landmarks

    d) objects

    e) buildings

3. Effective places are

a) clearly lit

b) similar to each other

c) very close to each other

d) very well-known

The pegword mnemonic

The pegword mnemonic is based on the same sort of idea as the place method, but instead of using locations as cues, it uses numbers. These numbers are transformed into visual images by means of the following simple rhyme:

One is a bun

two is a shoe

three is a tree

four is a door

five is a hive

six is sticks

seven is heaven

eight is a gate

nine is a vine (also, variously: nine is wine; nine is a line)

ten is a hen

One is a bun

Two is a shoe

Three is a tree

Four is a door

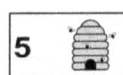
Five is a hive

Six is sticks

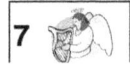
Seven is heaven

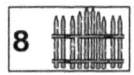
Eight is a gate

Nine is wine

Ten is a hen

The rhyme must be learned by rote until it is over-learned. Accordingly, the pegword method is not as quickly mastered as the place method, where cues already over-learned are used. It does however have an advantage over the place method, in that the items learned are not tied to a particular sequence, and therefore it's not necessary to recall the whole list to retrieve a single item. That is, you can simply ask yourself what number 7 is, without having to go through the first six items to get there.

It is also ideal for learning numbered lists, such as the cranial nerves:

1 is a bun and cranial nerve 1 is olfactory. So you could visualize a nose diving into a bun

2 is a shoe, and cranial nerve 2 is optic

3 is a tree, and cranial nerve 3 is oculomotor — so continuing with the eye theme, and bringing in something that cues us to motor

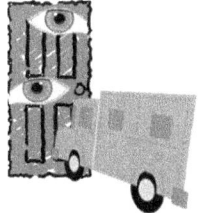

4 is a door, and cranial nerve 4 is trochlear — still with the eyes, and bringing in a truck to cue us to the name

The pegword mnemonic

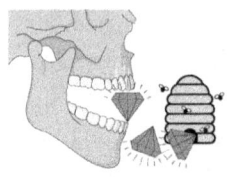 5 is a hive, and cranial nerve 5 is trigeminal, which relates to the jaw — so something to cue the jaw and something to cue the name (three gems)

6 is sticks, and cranial nerve 6 is abducens — back to the eyes, and bringing in a ufo to cue the name (alien abduction)

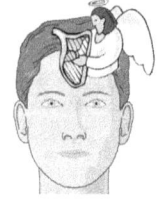 7 is heaven, and cranial nerve 7 is facial

8 is a gate, and cranial nerve 8 is auditory, so let's give our gate some ears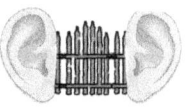

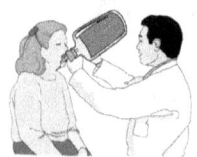 9 is wine, and cranial nerve 9 is glossopharyngeal, which relates to the throat

10 is a hen, and cranial nerve 10 is vagus, which relates to the heart, which we'll do in pale pink and blur the edges, to make it a little vague

The rhyming pegwords I gave you only went to 10, but we can add to this traditional list:

Eleven is a lemon   Twelve is shelves

Cranial nerve 11 is accessory, which relates to head movement, so we use a bag to indicate an accessory, and try to put our lemons and our shrunken head into it

Cranial nerve 12 is hypoglossal, which relates to the tongue, so we use a hypodermic needle to cue the name, and throw in the tongue and our shelves

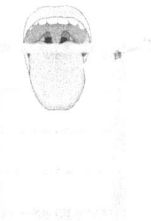

To return to the principle I discussed under the keyword method: images are effective to the extent to which the elements interact actively and meaningfully. For this reason, I found my seventh image poor. There is no particular connection between the face and the angel in this image. A better image would be an angel having a facial, perhaps with lurid green sludge over their face.

The pegword method can be considerably extended using the coding mnemonic. We'll look at that in the discussion of that mnemonic.

Applying the pegword method

One of the potential problems with the pegword method is the question of how many lists you can hang off the same pegs. However, one early study[1] found no problems with interference when volunteers learned six consecutive reorderings of the same nouns (although the lists did not have to be remembered very long, and I strongly suspect interference would increasingly have become a problem for memory over time).

Similarly, a recent study[2] found dramatic benefits from using a combined pegword-keyword mnemonic to remember three different lists (the 10 highest mountains, the 10 tallest waterfalls, the 10 largest volcanoes), as well as two training lists: tested five days after learning these, average recall was 63.5%, 51.3%, and 30.9% respectively for those using the mnemonic, compared to 42.2%, 26.1%, and 17.4% for those using their own methods.

In other words, yes there is interference, but using the pegword mnemonic is decidedly better than not using it. Nor do we know how much more effective the mnemonic may have been if the lists had been spaced out.

Interference is always going to occur if you try and learn different lists at the same time, regardless of method. I feel the best way is to space out your lists a little, achieving a certain level of mastery of one list before tackling a new one. If you do feel there might be some confusion between the lists, you might find it helpful to provide a distinctive mark for each list (that appears in the mnemonic image for each item on the list).

Although concrete, visualizable words are always better if you can find appropriate ones, two studies[3] have found abstract pegwords (e.g., one is fun, two is true, three is free, four is more, five is alive, six is for kicks, seven is even, eight is late, nine is fine, ten is then) could be just as effective as concrete ones. But this has not been found consistently$_4$. It seems to depend on your imagery ability$_5$.

> ### The pegword method
> - requires you to learn the pegs very very well
> - enables you to go directly to any item on a list
> - is the best method for a numbered list

Glossary

pegs — clavijas

sludge — lodo

volunteers — voluntarios

consecutive — consecutivo, siguiendo en secuencia

reorderings — reordenamientos

dramatic — dramático, grandes

respectively — respectivamente

regardless — independientemente

Review 8.1

1. The pegword method is the best mnemonic for
   a) linked pairs
   b) new words
   c) passwords
   d) numbered lists
   e) main points of a speech

2. Match the pegwords to their number

tree	3
bun	2
hen	9
sticks	8
shoe	1
wine	4
gate	10
door	5
hive	6

3. Which pegword is represented in this image?

a) angel
b) seven
c) face
d) heaven

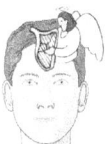

4. Which pegword is represented in this image?

a) skull
b) hive
c) four
d) five

5. This image uses what mnemonic strategies?

a) story mnemonic
b) place method
c) keyword method
d) pegword method

The link method

Like the place method and the pegword method, the link method uses visual images to link items together. However, instead of using a well-learned structure to anchor the new information, items are linked to each other. In this way it is like the story method.

However, the link method requires less thematic coherence than the story method — you are essentially building a chain, in which the only requirement is that each item forms a visual image with the item next to it.

Thus, for our 12 cranial nerves, you could form the following images:

a **nose** with the **eyes** above (glasses rather than eyes help distinguish the optic nerve from the oculomotor nerve)

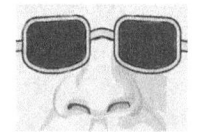

the **eyes** on the handlebars, riding a **motorbike** (here's the reason for using glasses — we need the eyes for the oculomotor nerve)

the **motorbike** hitting a **truck**

the **truck** spilling its load of **3 gems**

a **ufo** coming down, abducting the **truck**driver;

an alien (**abductor**) ripping off the truckdriver's **face**;

the **ears** falling off the **face**;

the **ears** falling onto a **glossy photo**;

the **glossy photo** pinned to a **pink heart**;

the **pink heart** being tucked into a bag (**accessory**);

the bag (**accessory**) being pierced with a **hypodermic**

Effectiveness of the link method

There has been very little research into the effectiveness of the link method, but findings from two early studies[1] found dramatic benefits from using this method to learn lists of unrelated nouns after a single presentation.

Bugelski, for example, found the average recall of a 20-word list without using the method was around 10 words, of which only half were remembered in the correct order, compared to half as many more being remembered using the link method, and markedly better order recall (in the most effective condition, in which college students were given 6 seconds to study each word, an average of 15 words were recalled in correct sequence, with an average of 17 recalled overall — this compares to an average of 5.5 and 9.7 when the students were simply told to try and learn the words in the order given).

Nor did the students receive much instruction in the technique (they were simply told to form an image of each word in their mind and to try to incorporate this image with the prior image in some interaction), suggesting that this is a method easily mastered — although like all mnemonics, practice will certainly make you much more fluent at implementing it.

The study also showed that the method had a marked effect on persistence and consistency.

Those using their own methods showed the typical pattern of remembering the first and last items best, with a sudden drop after the 3rd item (the primacy and recency effects). By the 7th item, these students had essentially given up. Those using the link method, however, simply showed a slow decline, with no precipitous drop. By the 7th item, 85% of the link method group were still learning.

Perhaps most importantly, the study found wide individual variability: around a sixth of the link method group (16 out of 96 students) recalled all 20 items, in sequence, after a single presentation. Some of these could, when asked, return all 20 in the reverse order.

This emphasizes how the effectiveness of a strategy depends as much on the individual as on the strategy itself.

> **Advantages & disadvantages of the link method**
>
> - doesn't require a pre-learned structure
> - is easily learned
> - doesn't allow you to go directly to any item, but requires you to work your way through the chain until you get to the item
> - requires visualization skill

Glossary

abducting — secuestrando

accessory — accesorio

hypodermic — hipodérmico

markedly better — notablemente mejor

prior — anterior

persistence — persistencia

consistency — consistencia

primacy — primacía

essentially given up — esencialmente abandonado

pre-learned — pre-aprendido

Review 9.1

1. The link method is the visual equivalent of

   a) an acrostic

   b) the story mnemonic

   c) the keyword method

   d) the first-letter mnemonic

2. The benefits of the link method are

   a) it doesn't require any skill at making visual images

   b) it's easily learned

   c) it allows you to go straight to any item in the sequence

   d) you don't need to memorize a framing structure, like a route of places, or numbered items

Part IV

Advanced mnemonics

Through this book, we have moved through the different mnemonic systems in terms of increasing difficulty. Having begun with the simplest, most familiar mnemonics, we moved to the keyword method, which is a powerful and effective technique in its own right, and also the component that turns the list-mnemonics into truly useful strategies.

In this final section, we will look at another mnemonic that is both very powerful in its own right, and a component that greatly expands the possibilities of the list-mnemonics.

Then I will bring all this information together, to help you use these mnemonics most effectively.

Coding mnemonics

A system for remembering numbers

Most people find numbers — phone numbers, personal identification numbers, dates, and so on — more difficult to remember than words. That is, of course, why businesses try to get phone numbers that correspond to some relevant word. The system whereby this is possible — the linking of certain letters to the different digits on a telephone calling pad — is a kind of coding mnemonic. Basically, coding mnemonics are systems that transform numbers into words.

Because words are much easier for most of us to remember, this is a good way to remember numbers, but it's not the only one. If you have a facility for numbers, or an existing store of memorized numbers (dates, baseball scores, running times, whatever), you can use those memorized numbers or your understanding of mathematical patterns to remember new numbers.

In one well-known experiment[1], for example, the subject was able (after 250 hours of practice!) to recall up to 82 digits after hearing them at the rate of one digit a second. This subject was a runner, and used his knowledge of record times to make the digit strings more memorable.

The difficulty with a coding system is that you can't use it effectively until you have fluently memorized the codes, to the extent that the linked letter (if encoding) or digit (if decoding) comes automatically to mind.

This requirement makes this sort of mnemonic the costliest of all the mnemonics — that is, it takes the most time and effort to master.

Of course, the pegword mnemonic is also a coding system, in a way — which is why it's harder to master than the method of loci. But the pegword mnemonic is easier than the digit-letter substitution mnemonic. There are two reasons why:

- the pegs and their numbers are connected by simple rhymes
- the construction of a composite image incorporating the peg and the item to be learned is less constrained than the finding of a suitable word constructed from the required letters.

Let me show you what I mean.

On the next page I have set out the best-known digit-letter code. It's important to note the system is based on sound rather than actual letters, so various similar-sounding letters are regarded as equivalent.

Now these substitutions weren't chosen arbitrarily. The inventor has tried his best to make them easy to remember. But as you can see, some of the rationales are somewhat contrived. If you think you can come up with codes that are easier for you to remember, feel free to change them — just bear in mind the dangers of confusability. For example, you could code *f* for *5*, but there is a strong risk of becoming confused between *4* and *5* when decoding.

> 0 = s, z, soft c (*zero* starts with a *s* sound)
>
> 1 = t, d, th (there's *1* downstroke in *t*)
>
> 2 = n (*2* downstrokes in *n*)
>
> 3 = m (*3* downstrokes in *m*)
>
> 4 = r (*r* is the last letter of *four*)
>
> 5 = l (*l* is *50* in Roman numbers)
>
> 6 = sh, ch, j, soft g (*six* has a sort of *sh* sound)
>
> 7 = k, q, hard g, hard c (number *7* is embedded in *k*)
>
> 8 = f, v (both 8 and *f* have two loops)
>
> 9 = p, b (*9* is *p* the wrong way round)

Once you've encoded the digits into letters, you can turn numbers into words or phrases or rhymes. Only consonants are used for coding. This means you can throw vowels (and also, in this system, *w*, *h* and *y*) in as necessary.

In this way the date 1945 could be encoded as *tprl*, which could be turned into *top role, to pour low, tie a poor owl, tip or lie, top rail*. Or *dprl: die poorly! tbrl: tuba role; dbrl: dab rule*.

You see what I mean about constraints. However, there is enough give in the system to make it possible to always come up with *something*.

If you do want to learn this particular system, there is a mnemonic that may help you memorize the 0-9 codes (from

Bower, 1978): **S**a**t**a**n m**ay **r**e**l**i**sh c**o**ff**ee **p**ie. *Why* will help you remember which consonants can be used freely, like vowels.

The system also allows you to use doubled consonants where the sound doesn't change (which is mostly). For example, *dipper, dabble, squirrel*. Compare these to *accent*, where the first *c* is hard (7) and the second *c* is soft (0).

Similarly, a silent consonant doesn't count. Thus *knee* equates to 2 and not 72. Less obviously, words like *would, could, should* don't count the *l*. But I'm not sure I like this myself. I'm very aware of how words are spelled, and tend to 'see' words as I hear them. I think whether or not you count silent consonants depends on your awareness of what the words look like.

A similar issue of personal preference is whether you regard *ng* as a variant of hard *g* (7), or two sounds: *n* and *g* (27).

x generally makes the sound of *ks* (70), but when it starts a word (xylophone, xenon) it's often more a soft *z* sound (0).

You can see why this system takes more training than the other systems! And why this mnemonic is the least studied of the major mnemonics. In those few studies that have been done, there are usually only one or two subjects. This is not surprising when you consider the number of hours needed to achieve mastery of this system. These studies have invariably focused on training their subjects to memorize very long strings of digits quickly. It's hard to imagine the everyday circumstances in which this would be a useful skill for most of us.

Having said that, there are a number of occasions when you want to remember shorter numbers — say four, or seven, or even nine digits. And the code really isn't as hard to learn as it might seem, looking at it. A little practice coding numbers into letters, and back again, works wonders in cementing this in your brain.

One study[2] that did involve a number of students, and, most interesting of all, did compare the performance of students who learnt the mnemonic with the performance of students instructed in general cognitive strategies (such as chunking and clustering) but not mnemonics, found impressive results with the mnemonic. Over 45% of the mnemonic students recalled all 20 digits in a 4x5 matrix, compared to 7% of the cognitive students. Over 85% recalled at least 16 digits, compared to 35%. Even more impressively, 43% recalled all 50 digits from a 50-digit matrix (4x12+1x2), and 78% recalled at least 45. The comparison group in this case were general psychology students who had received no particular cognitive training — none remembered more than 34 digits. Indeed, only 3 of the 37 mnemonic students who learned the 50-digit matrix did as badly as any of the general students, and those 3 had all failed to learn the mnemonic properly.

The mnemonic students had spent four 75-minute classes studying the mnemonic, plus about an hour's practice outside class. They studied (but didn't memorize) a list of 100, but didn't have much practice memorizing matrices (one 60-second practice with a 20-digit matrix).

> ### Keeping it simple
>
> Remember the codes with: **Satan may relish coffee pie**
>
> Vowels and consonants **why** don't count
>
> Avoid **x**, **ng**, silent consonants.
>
> Remember doubled consonants that sound like one only count as one
>
> Remember it's the sound that matters.

Extending the coding mnemonic with other mnemonics

The coding mnemonic on its own is simply a way of transforming poorly-remembered numbers into better-remembered words. But words and phrases, as we have seen, vary in their memorability. Studies$_3$ have shown that the difficulty of coming up with easily remembered words is sufficiently great that even with two-digit numbers, students don't easily come up with memorable words when time is limited. Thus, only when good words are supplied by the experimenter, does the method produce significantly better recall.

Of course, the key phrase here is *when time is limited*. In normal (non-experimental) circumstances, you will have as much time as you need to come up with good words and practice them sufficiently. But the limitation does point to the difficulty of the method for novices (it may well be, as Higbee suggests, that self-generated coded words are more effective when the mnemonist is more skilled).

It also points to the reason why lists of coded words are popular — however, those do require that you sit down and memorize them all! Whether or not that time and effort is worth it to you depends on how much you're going to use it. And that depends, in part, on *how* you are going to use it. I have talked of the coding mnemonic as a transformational strategy, but it is also a powerful tool as part of what Higbee calls a *mental filing system*.

But first, let's explore more deeply the use of the coding mnemonic in remembering numbers. Let's try the code out on some longer numbers.

237-812-469 = n-m-k-f-t-n-r-sh-p

Any meek feet now rush by.

Here's another:

3794-2106-6512-8843 = m-k-p-r-n-t-s-sh-sh-l-t-n-f-f-r-m

My keeper and Sasha shall wait in a fever home.

As you can see, the longer the numbers, the harder to make a memorable sentence! But I didn't cheat — no tweaking of the numbers (which I produced off the top of my head) in order to create a more meaningful sentence. On the other hand, you could probably come up with better sentences if you spent more time on it (I came up with these in just a few minutes).

Another way of dealing with long numbers is to break them into 2-digit groups and use a pre-learned list of 100 one- and two-digit words which you then convert into a series of linked images. Thus

237-812-469 = 23-78-12-46-9 —> gnome-coffee-tin-rash-pie

3794-2106-6512-8843 = 3-79-42-10-66-51-28-84-3 —> ham-cap-rain-toes-judge-wallet-knife-fire-ham

You'll notice that in the second example, even though the number would have broken evenly into pairs, I put a single digit at the beginning and the end (not ideal in this particular case since the number began and ended with the same digit). I did this because otherwise one of the pairs would have been 06. The other solution would have been to make a combined image (a witch (6) with a hose (0)).

If you prefer words to images, you could also make these words into a story rather than a chain of images. For example, the gnome pours his coffee into a tin, gets a rash, and solaces himself with a pie. Here you can see the importance of making sure your code-words are all nouns — that way you know that only the nouns in the story signify numbers.

There are other ways to organize long numbers. In the case of the runner who used track times, for example, the times provided small groups of digits, which the runner then organized into small clusters, and then into super-clusters.

Another strategy is to use the place method. In one study[4], for example, a subject memorized 100 concrete nouns to match the numbers 0 to 99, and 40 Berlin landmarks (in a particular order). They could then encode a string of 80 digits by mentally attaching the image for each digit pair to the landmark.

But it did take many hours to master this!

In another study[5], a subject combined the story mnemonic with the place method by making a story for each group of 10 words (20 digits — again, it was first necessary to memorize 100 concrete nouns for each possible digit-pair combination), and then placing each story in a different setting. These settings were of course always used in the same order. Using this combination resulted in fewer errors and recall failures than the story mnemonic on its own. Of course, as always with studies involving coding mnemonics, we are drawing conclusions from the performance of one or two people only, so we cannot be too firm about this! However, it does make sense.

Practical uses for coding mnemonics

In the study situation, there is quite a lot of numerical

information that you might need or be interested in learning. For example, historical dates, mathematical formulae, geographical facts.

Let's look at how we can combine three different mnemonics to remember the lengths of the ten longest rivers (in miles):

1. Nile (4140 miles)
2. Amazon (3990 miles)
3. Yangtse (3960 miles)
4. Yenisei-Angara (3445 miles)
5. Ob-Irtysh (3360 miles)
6. Hwang Ho (3005 miles)
7. Zaire/Congo (2900 miles)
8. Amur (2800 miles)
9. Mekong (2795 miles)
10. Lena (2730 miles)

First, let's use the coding mnemonic to convert the lengths:

1. Nile 4140 = r-t-r-s = rotors, raiders, readers
2. Amazon 3990 = m-p-p-s = my pipes
3. Yangtse 3960 = m-p-sh-s = impish ass
4. Yenisei-Angara 3445 = m-r-r-f = more or few, more rev
5. Ob-Irtysh 3360 = m-m-sh-s = mommy shoes
6. Hwang Ho 3005 = m-s-s-f = mass shave, miss safe
7. Zaire/Congo 2900 = n-p-s-s = any passes
8. Amur 2800 = n-f-s-s = no fusses

9. Mekong 2795 = n-k-p-f = neck puff

10. Lena 2730 = n-k-m-s = hen games

They're not brilliant, I know. Sometimes numbers will fall nicely into meaningful words, but more often than not they won't. Still, you'll be surprised how much these rather weird phrases help.

Now we need to use the keyword mnemonic to turn the names of the rivers into something concrete and familiar.

Nile — nail

Yangtse — ant sea (a sea of ants)

Yenisei-Angara — nice anchor (*nice* is not concrete, but you can attach it by always thinking *nice anchor* when visualizing it)

Ob-Irtysh — (observe the) yurt

Hwang Ho — hanging (image of a noose hanging from a gallows)

Zaire/Congo — stair (not using Congo because of its similarity to Mekong)

Amur — a mule

Mekong — King Kong

Lena — lion

You'll notice I didn't give a keyword for Amazon; I felt the female Amazon would provide an adequate image.

Now we can use the pegword mnemonic to provide our

ordered list. So we tie our keywords to the pegwords to produce the following images:

1. a nail in a bun
2. an Amazon with one big shoe
3. a sea of ants around a tree
4. an anchored door
5. someone watching a yurt with a beehive hanging from its pole
6. an axe among sticks next to the gallows
7. stairway to heaven
8. a mule nudging a gate
9. King Kong drinking wine
10. a lion ripping apart a hen

Once you've worked on visualizing these images and got them well down, you can then connect the images to your sentences. Don't worry if your images aren't as clear as you think they should be; I rely equally on the words as much as the images — but it helps to visualize as much as you can while thinking on the words.

Now you've got these word-images (and it really does take very little practice), you can stick the coded phrases on.

1. a nail in a bun: **rotors** (helicopter blades) trying to lift the nailed bun
2. an Amazon with one big shoe: tripping over her big shoe, she drops her blowpipes: **My pipes!** she cries
3. a sea of ants around a tree: an **impish ass** (donkey) grins as he flicks ants at the tree

4. an anchored door: the door strains to move against its anchor; I shout: **More rev!**
5. someone watching a yurt with a beehive hanging from its pole: the observer is wearing **mommy shoes**
6. an axe among sticks next to the gallows: a mass of people lining up to be shaved by the axe (**mass shave**)
7. stairway to heaven: "**any passes**?" I ask anxiously
8. mule nudging a gate: going through easily; "**no fusses**"
9. King Kong drinking wine: big **neck puff** around his neck
10. a lion ripping apart a hen: **hen games**!

It all sounds very strained and unnecessarily complicated if you simply read all this! You absolutely cannot appreciate this method until you try it. It really is much simpler than it appears (although still not a simple strategy). However, it is vital that you build up the strategy step by step. In this case, for example, you must be fully confident of the standard 1-10 pegwords (1 is a bun, etc) first; then you fix the rivers to the pegwords firmly; and finally you attach the coded phrases.

Notice also how my choice of phrases was modified by the pegword-river images. For example, originally I preferred 'raiders', but when I came to attach it to the nailed bun, I found it easier to integrate helicopter blades. Similarly, "more or few" made more sense to me until I considered it in relation to the anchored door, when 'more rev' fit in better.

Again, if you actually do this exercise and try to learn these images, you'll see for yourself how the easiest ones to learn are the ones that make more meaningful, more tightly integrated, word-pictures.

I remarked earlier on the awkwardness of some of these phrases. There is another way of dealing with the codes — you can make a sentence with each word in the sentence starting with the letter you want to remember.

Thus:

Rust **d**own **r**oof **s**lates.

Many **p**irates **p**ester **s**almon.

Monkeys **p**ick **s**hells **s**lowly.

Men **r**ead **r**ed **f**ables.

Monks **m**ine **sh**ell **s**ands.

May **s**ell **s**uper **f**eet.

None **p**ay **s**illy **s**ums.

Names **f**or **s**imple **s**ongs.

Never **k**ick **p**oor **f**ellows.

Nearly **k**isses **m**y **s**on.

There are two ways you could attach these to the rivers. You could do it in a similar way to the coded phrases:

- as you visualize your nail in a bun, you say *Rust down roof slates*
- as you picture your Amazon in her big shoe, you think: *Many pirates pester salmon*
- and so on.

(By the way, never get so hung up on specific mnemonics you forget to use other information. It does help if you pay attention to features such as, in this case, that only the first sentence starts with a different letter, an R, while the next five begin with M and the last four with N. In the same way, with the cranial nerves, it helps to note that there are three Os, two Ts and three As; that four of the nerves are to do with the eyes and two have *gloss* in the name. These can all be useful cues that will help trigger the mnemonic if you forget it.)

The other method of attaching the coded sentences to the river names is to keep the two mnemonics as completely separate lists, and incorporate the river keyword into the sentence:

(Nails) Rust down roof slates.

(With the Amazon) Many pirates pester salmon.

(In the ant sea) Monkeys pick shells slowly.

(At anchor) Men read red fables.

(In the yurt, observant) Monks mine shell sands.

(Hanged Whores) May sell super feet.

(On the stairs) None pay silly sums.

(Mule) Names for simple songs.

(King Kong) Never kicks poor fellows.

(A lion) Nearly kisses my son.

In this instance I think the coded phrases, however awkward they may sound, work better than these mnemonic sentences, but there are circumstances in which you may well find a sentence works better.

All of this sounds terribly complicated, I know. But the point is not that these mnemonics are simple — they aren't! Nor does the use of mnemonics mean you don't need to put in any effort; it's not a magic trick (or perhaps that is exactly what it is — magic tricks look very impressive to the audience, but the magician has had to practice hard to produce them).

The point of these mnemonics is to make raw facts such as these (facts which are not meaningfully connected) more memorable. And they do, although you probably won't believe it until you try it. But although the method I've described may seem very complicated, it really does make learning and remembering these facts much easier than the alternative: brute force (rote repetition).

Dealing with decimals

Higbee suggests[2] that you can use *s* to mark decimals. Because numbers will never start with 0, coded words will never begin with *s* (of course they may begin with *sh*). Thus you are free to use *s* as an initial to indicate a decimal point. For example:

389.75 = m-f-p s-k-l = move by skill (note the *s* must start the word on the other side of the decimal point; you can't just put it into a word, for otherwise it would indicate a 0)

3.14159 = m s-t-r-l-p = my stroll-by

0.469 = s-r-shp = sir sheep (note you don't need to encode the marker 0 before the decimal point)

Of course, if you do use this, it means you should never start a word with *s* when you are transforming a long number into several words, unless it is a decimal point.

Another way of doing it is to replace a decimal point with a comma in your phrase or sentence, but this works better if you're seeing the word as well as hearing it.

Retrieval

One of the reasons why coding mnemonics are more effortful is that it takes practice before you can automatically and fluently code numbers into letters.

The other reason is that it takes an equal amount of practice to *decode* automatically and fluently! In other words, it is not enough to practice turning numbers into words. To master this strategy, you need to also practice turning words into numbers.

I suggest that if you want to master this technique, you practice transforming odd phrases and sentences as you read the newspaper, or other casual material (like the back of the cereal box).

Other languages may work better!

While I have restricted my discussion to English, it shouldn't go unremarked that the coding mnemonic may be easier in some other languages. Japanese, for example, is a syllabic language, meaning they have a different sign for each consonant-vowel pair.

Moreover, because they have built their written language on

top of the characters imported from China, their digits are each associated with two or more different pronunciations. For these reasons, transforming numbers into letter-sounds is sufficiently easy and obvious that many Japanese spontaneously use such a mnemonic[6].

Using the coding system to extend the pegword mnemonic

As mentioned, the value of this code goes beyond simply creating words or phrases that help you remember numbers. You can also use it to create concrete images that can become pegs.

Let's look at possible words for various numbers:

1. **t**ie (man's necktie)
2. ho**n**ey (remember, h and y don't count!)
3. ha**m**
4. ea**r**
5. ow**l**
6. wi**tch**
7. ya**k**; e**gg**
8. **f**ae; i**v**y
9. **p**ie
10. **t**oe**s**
11. **t**oa**d**; **d**ea**th**
12. **t**i**n**

13. **t**i**m**e; **d**i**m**e
14. **d**ee**r**
15. **t**owe**l**
16. **d**i**sh**
17. **d**u**ck**
18. **d**o**v**e
19. **t**o**p**; **t**a**p**
20. **n**o**s**e

And so on, up to 99.

As you can readily appreciate, this enables you to greatly extend the pegword method.

Moreover, although finding concrete words becomes increasingly hard as the number of digits increases, you can use a kind of hierarchical organization to extend your pegwords without having to come up with appropriate words. For example, by tying colors or other adjectives to particular groups, so that the nouns for numbers 100-199 are all described by one particular adjective, those for 200-299 by another, and so on.

The use of pegwords in this way is clearly not aimed only at numbered lists (although it can be very useful if you want to memorize Bible verses). It is in fact the basis for a mental filing system that you can use to organize huge amounts of information, to make for easy retrieval.

However, there is no research or even anecdotal evidence of successful use of such large systems, and I believe it is only likely to be of any use for experts — those who are not only expert in mnemonics, but also in other, meaning-based, memory strategies.

So I do not recommend you dive into this particular system, but concentrate first on building your expertise in all the other (simpler!) strategies that help you learn and remember.

> **The coding mnemonic**
>
> - requires the most time to master
> - is an effective means of remembering numbers
> - can be combined with list-mnemonics to learn very long numbers
> - can be combined with the pegword method to extend the number of pegs

Glossary

correspond — corresponden

relevant — pertinente

system — sistema

whereby — por lo cual

facility — facilidad, talento

decoding — descodificación

automatically — automáticamente

costliest — más costoso

effort — esfuerzo

substitution — sustitución

composite — compuesto

equivalent — equivalente, lo mismo

downstroke — trazo que baja

contrived — artificial, forzado

bear in mind — tener en cuenta

enough give — suficiente flexibilidad

works wonders — funciona de maravilla

chunking — haciendo trozos

clustering — recogiendo en grupos significativos

impressive — impresionante

matrix — matriz

mnemonist — mnemonista, un experto en técnicas mnemotécnicas

tweaking — retocar, cambiar un poco

off the top of my head — sin pensarlo dos veces

solaces — consola

setting — ajuste

weird — extraño

noose — soga

gallows — horca

awkwardness — torpeza

complicated — complicado

magician — ilusionista

brute force — fuerza bruta

comma — coma

owl — búho

witch — bruja
fae — hada
ivy — hiedra
anecdotal — anecdótico
concentrate — concentres

Review 10.1

1. The coding mnemonic is for remembering

    a) simple facts

    b) dates

    c) new words

    d) any numerical information

    The next questions all relate to the digit-letter code. Try to pick the correct answer as quickly as you can — speed counts, because to use this mnemonic effectively, you need to know the basic code as well as you know your letters and numbers.

2. Pick the number paired with the letter: n

    a) 5

    b) 8

    c) 2

    d) 3

3. with the letter: k
   a) 6
   b) 8
   c) 3
   d) 7

4. with the letter: t
   a) 1
   b) 2
   c) 9
   d) 3

5. with the letter: p
   a) 8
   b) 9
   c) 0
   d) 6

6. with the letter: s
   a) 3
   b) 0
   c) 7
   d) 6

7. Now pick the letter paired with the number: 8

   a) f

   b) g

   c) b

   d) m

8. with the number: 5

   a) q

   b) l

   c) v

   d) f

9. with the number: 3

   a) n

   b) m

   c) w

   d) r

10. with the number: 4

   a) r

   b) d

   c) s

   d) f

11. Pick the correct decoded number for the word: knee
   a) 34
   b) 72
   c) 73
   d) 2

12. Pick the correct decoded number for the word: accent
   a) 7021
   b) 721
   c) 71
   d) 621

13. Pick the correct encoded word for the number 3412
   a) march
   b) martin
   c) wafts
   d) sifter

14. Pick the correct encoded phrase for the number 231-810-501
   a) never stand down
   b) mind sits fast
   c) down that task
   d) named fates lost

15. The coding mnemonic can be used with the

   a) method of loci

   b) story mnemonic

   c) pegword mnemonic

   d) keyword mnemonic

Mastering mnemonics

What mnemonics are good for

Before we get on to choosing the right mnemonic strategy, let's briefly talk a little more about when mnemonic strategies are the best choice.

Assessing the text and the task

As I discuss in my book *Effective notetaking*, the first step in choosing the right strategy for a task is to accurately analyze the problem — indeed, to identify that there *is* a problem. One of the main reasons poor learners are less successful is because they are often unaware that they haven't understood something, or that they won't remember something without deliberate effort.

So, your first task is to recognize when a deliberate effort is needed. Then you need to accurately define the problem so that you know which of all the tools you should have in your toolbox is the right one for the task.

Defining the problem involves three actions:

- articulating your goal
- defining the retrieval context
- evaluating the material

I cannot emphasize enough that the first step in any memory task is to clearly specify your goal — at all levels, because goals are nested. For example, you may have the overall goal of getting a degree, the more specific goal of passing a particular exam, and the still more specific goal of understanding a particular chapter, or of committing to memory the elements in the periodic table. To accurately define these more specific goals, it helps to understand the wider goals they're nested in. To learn the "right" material — the material you need to learn to fulfil these goals— you need to have these goals clearly articulated.

Part of articulating that goal is realizing the circumstances in which you are going to be retrieving the material. For example, do you need it for an exam? for multi-choice questions or essay answers? for a classroom discussion? for an oral presentation? to build expertise?

Different situations make different demands on you. You may need to:

- **select** the relevant information;
- **organize** the relevant information into a clear structure;
- **understand** the information;
- **recognize** the correct information;
- **respond** with the right information when given a cue;
- **retrieve** the right information when needed.

My book *Effective notetaking* discusses in great detail the first three tasks (selection, organization, comprehension) — all of which, if done with sufficient attention, should be sufficient for recognition, and may well be sufficient for recall. Mnemonics of course is only about recall — retrieving information in response to some cue or when needed.

So, given that you've identified the need for retrieval, how do you know if you need mnemonics?

There are two parts to this answer. The first depends on the information itself, on the degree to which it connects with other information you know well, and whether or not it needs to be remembered in essence or with precision (a particular order; specific words).

The second depends on you and the retrieval context, on how likely you are to be able to retrieve the information in the situation you will be in. If you suffer from memory blocks when anxious or stressed, for example, mnemonics are a very good way of providing an additional access point to the information.

The bottom line is:

Mnemonics are the best type of strategy for cueing well-learned information (for overcoming memory blocks or for reminding you of the order of information or the names of things).

Mnemonics are the best strategy for memorizing arbitrary details.

Mnemonics can provide a retrieval structure (a framework to help you remember) when you lack the knowledge and understanding to use comprehension strategies such as elaborative interrogation or self-explanation.

Choosing the right strategy for the task

Given that you've decided that the situation calls for a mnemonic strategy, how do you decide which one is the best? There are three main factors you should consider:

- Information attributes
- Retrieval requirements
- Personal characteristics

Let's look at information attributes first.

As I've said, mnemonics are primarily for meaningless (which means unconnected) information. However, meaningfulness is not an either/or attribute. It is a product of the interaction between the individual and the information, and it varies across individuals, across time (what is meaningless for you at the beginning of your studies will hopefully not be meaningless for you by the end of them!), and across context.

Moreover, there are different types of meaningfulness. While deep meaning, and understanding, comes from connection to your existing knowledge, a certain level of meaning comes simply from internal relatedness — connection between the bits of information being presented to you.

Different degrees of this internal relatedness lend themselves to different mnemonic strategies. Thus at one end of the scale, rhythm and rhyme require a relatively high degree of relatedness, while at the other end, the place method requires no relatedness at all.

In general, the easiest and most common mnemonic strategies are also the ones that require a reasonable degree of relatedness:

- First-letter mnemonics

- Rhythm, rhyme, and song
- Story method

Although as a general rule the imagery-based list-mnemonics don't require much internal relatedness, the three different imagery list-learning strategies do vary in terms of the relatedness they require between the list items. Some items, for example, are linked by being sequential (ordered by number) — these obviously lend themselves to the pegword method. Others will be linked by being consequential (connected by cause and effect) — these lend themselves to the link method. Some items (particularly in experimental studies) may be arbitrarily related — this is the case where the place method is likely to be the best choice.

You'll recall also that the place method is particularly recommended in oral situations (such as when listening to lectures). Directly encoding information into a mnemonic while listening to a lecture is of course a job for a skilled practitioner! However, it certainly has its advantages if you do get sufficiently skilled, and in such cases a visual mnemonic seems to be better than a verbal one.

This idea that visual mnemonics are better for situations where you hear the information, and verbal mnemonics for situations where you see the information (as in written text) is a useful guideline, but comes with plenty of addendums.

For example, there's some evidence[1] that using the pegword method is less disrupted than the place method by doing another visual task at the same time, because while it is visual, it is not *spatial*, as the place method is (which thus puts more demands on your visuospatial pathway).

It's also speculated[2] that in the case of written texts we need to consider the content. Descriptions of visual scenes should be regarded as visuospatial; narratives as causal or thematic sequences of events in time; expository texts as following some logical process.

If this is so, it may be that the place method is best used on expository texts, the link method on narratives, and the pegword method on static visual descriptions.

In a similar vein, it has been suggested that the story method is better for abstract items, and imagery methods for concrete items. This presupposes that you're not using the keyword method to transform an abstract or difficult word into something more concrete and memorable, or that the best transformations are in fact abstract words.

Remember that list-mnemonics are not restricted to actual lists, but are useful for any ordered information. In the case of numbered lists, of course, the pegword method (possibly in conjunction with the coding mnemonic) is the best tool. But again, you should think of 'numbered' in broad terms: if you want to be able to go directly to any item on the list, and not be constrained by having to work through it item-by-item, you are better with a numbered list.

A major drawback of the place and link methods is of course that, if you want a specific item, you must start from the beginning of the list and work your way through it until you reach the item. This reinforces the matching of these methods to information that is sequential or consequential.

Both the place and the pegword methods share the property that they require a pre-learned structure. This property has the advantage of providing a retrieval structure, but also the disadvantage that — because the structure is being re-used — the most recent list is the one most easily recalled and earlier lists are recalled with more difficulty$_3$. You can of course reduce interference by giving each list its own special tag, or by using different places and pegwords. However, as with the keyword method, these techniques are best regarded as ways to enable you to quickly learn difficult-to-learn items — a learning which requires consolidation through retrieval practice.

A list summarizing the tasks for which each mnemonic is most effective appears at the end of this chapter. You should always remember the general principle, however: **first and foremost, the purpose of mnemonics is to provide you with retrieval cues**. Different mnemonics do that in different ways, but how effective they are depends not on the specific strategy but on how well that retrieval cue calls forth the information you need to know. That is the measure of your success, and it is according to that goal that you should determine which particular mnemonic will serve your purpose in any specific situation. So you always need to think about the circumstances in which you will be trying to retrieve the information.

Choosing the strategies that are right for you

Finally, of course, these guidelines need to be considered in relation to your own abilities. It's been noted[4] that people with extraordinary memories show no uniformity in the strategies they use. In other words, there is no 'magic bullet', no single path to expertise. Some do it through their powerful imagery ability, some through their painfully mastered mnemonic expertise, others through their mastery of organizational strategies. The recipes are all different, and this underscores the fact that there are many effective memory strategies.

It's also been suggested[5] that this variety of techniques reflects the fact that all these highly skilled memorists have developed their skills on their own. This emphasizes the personal component — the importance of tailoring effective strategies to your own strengths and weaknesses, and with an eye to what you enjoy and find motivating.

Successful strategies need practice

No matter what strategy you use to remember something, you're going to need a certain amount of repetition to fix it in your brain. Mnemonics allow you to shortcut the process — the measure of how 'good' a particular mnemonic is, is indeed the degree to which it reduces the need for repetition — but no mnemonic is going to do away with the need for some repetition.

But repetition is not as straightforward a strategy as it might appear. We all know how to repeat something, but a lot of that repetition is wasted effort. The basic principle of effective repetition is quite simple: **effective repetition occurs at increasingly spaced intervals**.

How far apart should they be? Well, you should start by testing your memory after a very brief interval — perhaps an hour. If you're successful with that and the memory is fluent (easily recalled), you could leave the next test until bedtime — I say that not because of the time factor, but because sleep consolidates what you've learned that day, and I'm a big believer in refreshing your mind with that information just before sleep. If all goes well, you could leave it for 2 or 3 days, and then for a week.

How far you extend that depends on how long you need to remember the information for. If you want to remember it permanently, you should review it after two months, and probably again after six months. That's based on a recent large study$_6$ that explored the optimal gap between study and test: to remember for a week, the optimal gap was one day; for a month, it was 11 days, for 2 months (70 days) it was 3 weeks, and similarly for remembering for a year.

That's just to give you some idea — it really does depend on the individual, so don't take it as gospel! Moreover, it only looked at a very simplified learning situation and a single review.

The important principle is that you aim to find the maximum interval at which you can reliably recall the item. So if your memory fails, shorten the interval; if you succeed but with difficulty, keep the same interval; if you succeed easily, lengthen it. BUT: don't be fooled by early fluency into giving up your repetitions! The biggest reason for memory failure is people believing that being able to recall something easily means that they've learned it. Fluent recall is certainly a good signal if it happens after a lengthy period, say six months, but not if it happens within hours or a few days.

Of course, if the material you're trying to learn is something that 'naturally' comes up again and again (as most basic information will in the course of learning a subject), then you don't need to worry about this so much.

The other point that needs emphasizing is that by practice I mean **retrieval practice**, not simple rehearsal. In other words, you need to practice recalling the information, not just repeating it. The big advantage of mnemonics is that it gives you the help you need in the early stages to enable you to practice retrieval. It also reduces the need for quite so much retrieval practice. But never forget that retrieval practice is the most important part of learning.

And don't forget the importance of matching your retrieval

practice to the retrieval contexts you expect! If you're going to need to know the capital given the country, you want to practice retrieving the capital from the country; if you're going to need to remember the second line of a verse after saying the first line, then you want to practice retrieving the second line keyword/image in response to the first line keyword/image.

The final point about practice that I want to make concerns distributing. Distributing your practice doesn't involve only spacing it out. It also involves interleaving it with other practice. A study[7] that disentangled interleaving and spacing found that, when spacing was held constant, interleaving more than *doubled* test scores (77% vs 38%). This suggests that practicing different things is an effective way of improving your learning when time is limited.

It's also worth noting that the improved test scores from interleaving occurred despite the fact that performance was in fact poorer during practice. This brings us back to the perils of quick and easy fluency, and the notion of **desirable difficulty** (difficulties that the student can handle successfully, that engages processes that support learning[8]).

> ### Principles of effective practice
>
> Practice **retrieval**.
>
> **Space** your practice at **increasing intervals**: aim for the time when you are just able to remember.
>
> **Interleave** your practice with other topics.

A Final word:

Never forget that being a successful student is far more about being a **smart user of effective strategies** than about being "smart". More important than intelligence, more important than the number of hours you put in, the most important ingredient in your success is knowing good strategies and knowing when and when not to use them. Mnemonics are just one type of learning strategy that you should have in your toolbox, and although they can be dramatically effective, they can't be the only skill you have. In particular, the effective use of mnemonics requires good selection skills — learning will only take you so far if you don't know what information you should be learning!

So, remember to clearly state your goals. Don't neglect developing your selection and notetaking skills. Never lose sight of the fact that successful learning is all about being able to retrieve the right information when it's needed, so think about the retrieval context when you're learning. Remember that even the best mnemonics require effective practice.

And if your confidence wavers, recite the mantra:

It's about knowing how, knowing when.

You know how, and when. Now you need to practice. In the final section, we'll explore in great detail how to approach the learning of a scientific topic.

Glossary

toolbox — caja de herramientas

articulating — articulando, expresando claramente

evaluating — evaluando

nested — anidado

addendums — apéndices
disrupted — interrumpido
spatial — espacial
speculated — especulado
logical — lógico
presupposes — presupone
reinforces — refuerza
consolidation — consolidación
foremost — principal, primero
uniformity — uniformidad
underscores — subraya
shortcut — atajo
do away with — eliminar
straightforward — sencillo
wasted effort — esfuerzo inútil
permanently — permanentemente
optimal — óptima
gospel — evangelio
distributing — distribución
interleaving — intercalar
disentangled — desvinculados
perils — peligros
wavers — tambalea

Review 11.1

1. Mnemonics are the best strategy when

    a) you need to understand complex material

    b) you have a clear goal

    c) you don't know enough to build a meaningful retrieval structure

    d) you have a lot of related details

    e) you just need reminders of information you actually know well-known

2. To choose the best learning strategy, you need to

    a) work out your goal

    b) assess your relevant knowledge

    c) assess the degree to which the material is meaningful

    d) know what information needs to be memorized

    e) assess the degree to which the information is connected

3. Mnemonic strategies are sufficient for learning your subjects   T / F

4. Successful study requires a mix of strategies   T / F

5. A good mix of study strategies would include
   a) notetaking strategies
   b) practice strategies
   c) mnemonic strategies
   d) self-regulation strategies
   e) comprehension strategies

6. Mnemonic strategies
   a) don't need any practice
   b) are good for long-term learning
   c) are good for fast learning
   d) are best used sparingly

Summary of Mnemonic Strategies

Mnemonic	Tasks where it is most effective
Acronym	When the information is well-known to you, AND The order is important, AND The initial letters happen to fall into a word or pseudo-word
Acrostic	When the information is well-known to you, AND The order is important
Rhythm & rhyme	When you have small amounts of information that can be expressed in simple terms
Keyword	When you need to remember words or terms that have no obvious connection with words you already know, OR When you need to remember pairs of associated items that have no meaningful connection (such as capitals of states)
Face-name variant	When you wish to remember names and their associated images (such as artists' paintings, or types of animal)

Story	Especially for written text involving abstract terms, where items are linked sequentially or consequentially
Link	Especially for information heard, where items are concrete and linked sequentially or consequentially (as in narrative)
Place	Especially for information heard, where items are concrete and the text is expository
Pegword	Especially when you have a numbered list, OR Where items are concrete and the text is descriptive, OR When you have many items and need to be able to go directly to any item on the list
Coding	When you need to remember numbers, OR You need to extend a numbered list

Case Study

In the following case study, I explore in depth the issue of learning the geological time scale — names, dates, and defining events. The emphasis is on developing mnemonics, of course, but an important part of the discussion concerns when and when not to use mnemonics, and how to decide.

Now I realize you probably have no particular interest in learning the geological time scale, but I hope you will nevertheless study this material rather than merely read it. Like many things, you will only fully benefit if you engage with it!

I have gone into the learning process in what may seem at times overwhelming detail, and some of you may find the level of detail completely unnecessary, but others will find it very helpful. I cannot tailor the material to every reader's level of expertise!

The point of this exercise is to show you how to engage with your learning material — what sort of things you should think about, how to decide what to learn and how to learn it.

So let's begin.

The Geological Time Scale

Phanerozoic Eon 542 mya—present
Cenozoic Era 65 mya—present
  Neogene Period 23 mya—present
    Holocene Epoch 8000 ya—present
    Pleistocene Epoch 1.8 mya—8000ya
    Pliocene Epoch 5.3 mya—1.8 mya
    Miocene Epoch 23 mya—5.3 mya
  Paleogene Period 65 mya—23 mya
    Oligocene Epoch 34 mya—23 mya
    Eocene Epoch 56 mya—34 mya
    Paleocene Epoch 65 mya—56 mya
Mesozoic Era 250 mya—65 mya
  Cretaceous Period 145 mya—65 mya
  Jurassic Period 200 mya—145 mya
  Triassic Period 250 mya—200 mya
Paleozoic Era 542 mya—250 mya
  Permian Period 300 mya—250 mya
  Carboniferous Period 360 mya—300 mya
  Devonian Period 416mya—360 mya
  Silurian Period 444 mya—416 mya
  Ordovician Period 488 mya—444 mya
  Cambrian Period 542mya—488 mya

Precambrian 4560 mya—542 mya
  Proterozoic Eon 2500 mya—542 mya
  Archean Eon 3800 mya—2500 mya
  Hadean Eon 4560 mya—3800 mya

How do we set about learning all this? Let's look at our possible strategies.

Memorizing new words, lists and dates

Acronyms

A common trick to help remember the geological time scale is to use a first-letter acronym, such as the classic:

> **Cam**els **O**ften **Si**t **D**own **Car**efully; **Per**haps **T**heir **J**oints **Cre**ak? **P**ersistent **E**arly **O**iling **M**ight **P**revent **P**ermanent **R**heumatism.

(This begins with the Cambrian Period and moves forward in time; note that in this traditional mnemonic the Holocene Epoch is here thought of by its older name of "Recent Epoch".)

What's the problem with this, as a way of remembering the geological scale?

It assumes we already know the names.

The principal (and often, only) purpose of an acronym is to remind you of the *order* of items that you already know.

A common problem with acronyms (first-letter by definition) is that there are often repeats of initials, causing confusion. A more useful strategy (though far more difficult) might be to use the first two or preferably three letters of the words. This not only distinguishes more clearly between items, but also provides a much better cue for items that are not hugely familiar. For example, here's one I came up with for the geological time-scale:

Hollow **Ple**adings **Pli**ght **Mio**sis;

Olive **Eo**ns **Pal**l **Cre**ation; (or **Oli**ve **Eo**ns **Pal**m **Cre**dulous, for a slight rhyme)

Juries **Tri**ck **Per**plexed **Car**ousers;

Devils **Sil**ence **Ord**ered **Cam**pers.

Because it is extremely difficult to make a meaningful sentence with these restraints (largely because of rare combinations such as Eo- and Mio- and to a lesser extent, Pli, Oli, and Jur), I have used rhythm to group it into a verse. There's a slight rhyme, but it's amazing how much power rhythm has to facilitate memory on its own.

It is easier, of course, to construct a sentence with these items if you are allowed to include a few "insignificant" words (i.e., not nouns or verbs) to hold them all together. Here's a possible sentence, this time starting from the oldest and moving forward to the most recent:

Campers **Ord**er **Sil**ver **Dev**ils to **Car**ry **Per**sons **Tri**cking **Jur**isprudent **Cre**tins in **Pal**my **Eo**ns of **Oli**ve **Mi**lk and **Pli**ant **Ple**adings for **Hol**idays

The problem with both this and the "verse" is that they are too long, given their difficulty, to be readily memorable. The answer to this is organization, and later we'll discuss how to use organization to reduce the mnemonic burden. But first, let's deal with another problem.

Although the use of three-letter acronyms lessens the need for such deep familiarity with the items to be learned, you do still need to know the items. With names as strange as the ones used in the geological time-scale, the best strategy is probably the keyword mnemonic (or at least a simplified version).

Looking for meaning

But let's start by considering the origin of the names. If they're meaningful, if there is a logic to the naming that we can follow, our task will be made incomparably easier.

Unfortunately, in this case there's not a lot of logic to the naming. Some of the periods are named after geographical areas where rocks from this period are common, or where they were first found — these are probably the easiest to learn. The epochs in particular, however, are problematic, as they are very similar, being based on ancient Greek (in which few students are now trained), and, most importantly of all, being essentially meaningless.

Let's look at them in detail. The common cene ending comes from the Greek for new (*ceno*).

- **Holocene** is from *holos* meaning entire
- **Pleistocene** is from *pleistos* meaning most
- **Pliocene** is from *pleion* meaning more
- **Miocene** is from *meion* meaning less
- **Oligocene** is from *oligos* meaning little, few

Case Study

- **Eocene** is from *eos* meaning dawn
- **Paleocene** is from *palaois* meaning old

So we have

- Holocene: entire new
- Pleistocene: most new
- Pliocene: more new
- Miocene: less new
- Oligocene: little new
- Eocene: dawn new
- Paleocene: old new

You *could* find this helpful (remember that we're moving backward in time, so that the Holocene is indeed the newest of these, and the Paleocene is the oldest), but the naming is really too arbitrary and meaningless to be of great help.

Better to come up with associations that have more meaning, even if that meaning is imposed by you. Here's some words you could use:

- Holocene: holy; hollow; hologram; holly
- Pleistocene: plasticine; plastic
- Pliocene: pliable; pliant; pliers
- Miocene: my; milo; myopic
- Oligocene: oligarchy; olive; oliphaunt (! Notice that the words don't have to be familiar to the whole world, even the dictionary-makers; the important thing is that they have significance to you)
- Eocene: eon; enzyme; obscene (note that it is not

necessary for the word to begin with the same letter(s) — a particularly difficult task in this instance; what's important is whether the word will serve as a good link for you)

- Paleocene: palace; palatial; paleolithic

To tie your chosen word to the word to be learned, you must form an association (that's why it's so important to choose a word that's good *for you* — associations are very personal). For example, you could say:

- **Holograms are very recent** (the Holocene is the most recent epoch)

- **Glaciers are plastic** or **My glaciers are made of plasticine** (the Pleistocene was the time of the "Great Ice Age")

- **The pliant Americas joined together** or **Pliable hominids arose** (Hominidae began in the Pliocene, and North and South America joined up)

- **Mild weather saw Africa collide with Asia** (the Miocene was warmer than the preceding epoch; during this time Africa finally connected to Eurasia)

- **Elephants become oligarchs!** (during the Oligocene mammals became the dominant vertebrates)

- **Continents obscenely separate** (Laurasia, the northern supercontinent, began to break up at the beginning of the Eocene; Gondwanaland, the southern supercontinent, continued its breakup)

- **Pale from the disaster, we pull ourselves together** (the Paleocene marks the beginning of a new era, after the K-T boundary event (thought by many to be an asteroid impact) in which the dinosaurs and so much other life died)

Now this is not, of course, in strict accordance with the keyword method. According to this method, we should choose a word as phonetically similar to the word-to-be-learned as possible, and as concrete as possible, and then form a visual image connecting the two. While this is fine with learning a different language (the most common use for the keyword method, and the one for which it was originally designed), it is clearly very difficult to create an image for something as abstract and difficult to visualize as a period of time.

It's also often difficult to find keywords that are both phonetically similar and concrete. We must improvise as best we may. What you need to bear in mind is that you are searching for an association that will stick in your mind, and link the unfamiliar (the information you are learning) to the familiar (information already well established in your mind).

With this in mind, look again at the suggested associations. This time, think in terms of whether you can make a picture in your mind.

Instead of "Holograms are very recent", you might want to form an image of someone falling into a hole (tying the Holocene to the "Age of Humans").

Holocene

Pleistocene

Glaciers made of plasticine might stand.

If you can visualize very limber (perhaps in distorted postures) ape-like humans, Pliable hominids might be satisfactory, or you may need to fall back on the pliers — perhaps an image of pliers bringing North and South America together.

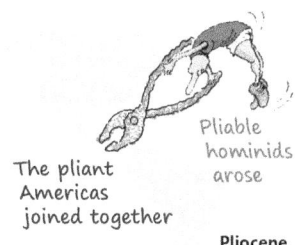

The pliant Americas joined together

Pliable hominids arose

Pliocene

Mild weather isn't terribly imageable; you might like to imagine milk pouring from the joint where Africa and Eurasia have collided.

Milk spills when continents clash

Miocene

Oligarchs is likewise difficult, but you could visualize elephants under olive trees, eating the olives.

Oligocene

Elephants love olives!

And now of course, we come to the most difficult — the Eocene. Here's a thought, for those brought up with Winnie the Pooh. If you have a clear picture of Eeyore, you could use him in this image. Perhaps Eeyore is standing on one part of the separating Laurasia (looking appropriately disconsolate).

Eeyore cries that the continents are parting

Eocene

Case Study

The Paleocene might best be associated with a palace, if we're looking for something imageable — perhaps dinosaurs sheltering in a palace as the asteroid comes down and destroys it.

Paleocene

You see from this that the demands of visual associations are often quite different from those of verbal associations. Both are effective. Whether you use verbal or visual associations should depend not only on your personal preference (some people find one easier, and some the other), but also on what the material best affords — that is, what is easiest, what comes more readily to mind, and also, which association will be less easily forgotten.

But mnemonics only take you so far. While very useful for learning new words, and for learning lists, they are not a good basis for developing an understanding of a subject — and unlike the situation of learning a language, a scientific topic definitely requires a more holistic approach. Mnemonics here are very much an adjunct strategy, not a complete solution. So before using mnemonics to fix specific hard-to-remember details in my brain, I would begin by organizing the information to be learned, with the goal of cutting it into meaningful chunks.

Glossary

in depth — a fondo

issue — tema

geological time scale — escala de tiempo geológica

defining event — evento que define una era

emphasis — énfasis

restraints — restricciones

slight — leve

power — poder

burden — carga

lessens — disminuye

simplified — simplificado

in strict accordance — en estricto acuerdo

phonetically — fonéticamente

limber — ágil

pliable — flexible

pliers — alicates

terribly — muy

collided — colisionado

likewise — de la misma manera

disconsolate — desconsolado

shelter — abrigo

destroys — destruye

demands — demandas, necesidades

readily to mind — viene a la mente de inmediato

holistic — holístico, tratado como un todo

Chunking your information

The first thing to note is that geological time is divided into two sections: the **Phanerozoic Eon** and the **Precambrian**. The crucial date is 542 million years ago. The Precambrian covers the period from the beginning of the earth up until 542 million years ago, and the Phanerozoic Eon covers the period from 542 million years ago until the present day.

So we have three important "facts": two names and a date. "Precambrian" simply means it is before the **Cambrian Period**, which is the first period of the Phanerozoic Eon; therefore, when you learn "Cambrian", "Precambrian" should slot easily into place. Let's begin with the more recent, and more complex, Phanerozoic Eon.

How should we remember this rather bizarre word? Let's start by considering its meaning. Phanerozoic is derived from the Greek *phaneros*, meaning "visible", and Greek *zoikos*, meaning "of animals". (The ending –zoic therefore indicates that the word relates to a specific manner of animal existence, and it's worth remembering that association because the suffix –zoic is used several times in geological time periods.)

The period was given this name because fossils of complex life become abundant. However, the name is less apt than it was, because many Precambrian fossils have now been found, indicating complex multicellular life.

So none of this is hugely helpful, and yet it can be used. You are likely to know that this most recent half-billion years is a time of abundant life. Noting the connection between –zoic and zoo might help you remember that –zoic indicates animals. You could remember the Greek word *phaneros* by noting its connection to "phantom", which is an illusion, a mere image. Thus, "Phanerozoic" — "visible animals", meaning we've found lots of fossils; from *phaneros*, visible, like a

phantom, and –zoic, meaning animals, as in a zoo.

Our visible animals need to be tied to the date of their beginning, and perhaps too, to an event. (Adding to the information you need to remember doesn't necessarily make your task greater; adding information can, paradoxically, often make your task easier — this is because meaningful information is always better remembered than meaningless.) The Phanerozoic Eon begins with the **Cambrian Explosion** — an unparalleled explosion of new life-forms.

This far back in time, dates are tricky things. The Cambrian Explosion was dated as beginning some 570 million years ago, but now it's thought to date from around 542 million years — this is why you'll see both dates cited.

Why Cambrian? It's named after Cambria, the Roman name for Wales, where rocks of this age were first studied. Cambria is the Latinized form of the Welsh name for their country: *Cymru*.

I mention all this not because it's necessary to know it, but (a) to make the information more meaningful, and (b) because any of these facts might connect with prior knowledge you possess. If any of this information is familiar to you, you can tie it to the new knowledge. If not, you might find it sufficiently interesting to be memorable. If it's not, I suggest you find a mnemonic that works for you (possible keywords include camber; cambric; camp; Camembert). There are also

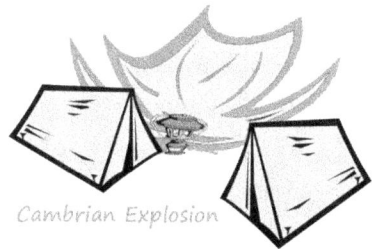

Cambrian Explosion

several places in the United States called Cambria; if you're familiar with one of those, it might serve as a connection.

Using a Coding Mnemonic to remember dates

Because there are a number of dates to memorize, you may well decide it's worth expending time on learning the coding mnemonic. Here it is again:

> 0 = s, z, soft c (*zero* starts with a *s* sound)
>
> 1 = t, d, th (there's *1* downstroke in *t*)
>
> 2 = n (*2* downstrokes in *n*)
>
> 3 = m (*3* downstrokes in *m*)
>
> 4 = r (*r* is the last letter of *four*)
>
> 5 = l (*l* is *50* in Roman numbers)
>
> 6 = sh, ch, j, soft g (*six* has a sort of *sh* sound)
>
> 7 = k, q, hard g, hard c (number *7* is embedded in *k*)
>
> 8 = f, v (both 8 and *f* have two loops)
>
> 9 = p, b (*9* is *p* the wrong way round)

According to this coding system, 542 would be lrn. You will remember that the system employs consonants only; that is to enable vowels to be used as required. Thus lrn could be rendered as learn, for example.

Having transformed the number into a word, you can incorporate that into a mnemonic with the associated information — the Cambrian Explosion. Perhaps a verbal mnemonic: **The Welsh learn trilobites explode in camp** (the first appearance of trilobites is one of the defining events for the beginning of this Explosion). You could construct a visual image of a camp flying the Welsh red dragon flag exploding with trilobites, with a teacher saying "Learn this".

But do we need a coding mnemonic to make this information memorable? Well, that is very much an individual decision, but it's premature to make it yet. Let's see how far organizing takes us.

Glossary

chunks — trozos

suffix — sufijo

phantom — fantasma

paradoxically — paradójicamente

unparalleled explosion — explosión sin precedentes

tricky — complicado

cited — citado

embedded — se incrustado

employs — emplea

More chunking

Let's reiterate what we know so far: geological time is divided into two eons, which are separated by a boundary marked by the Cambrian Explosion 542 m.y. ago. The first period — the Precambrian Eon — lasted from the birth of earth, some 4 ½ billion years ago, until the Cambrian explosion, approximately ½ billion years ago (meaning it lasted some 4 billion years). The second period — the Phanerozoic Eon ("visible animals") — lasted from 542 m.y. ago to the present.

[Precambrian Eon | Phanerozoic Eon]

Cambrian Explosion
542 m.y.

See this image? Doesn't it have a much greater impact on your understanding than that dry recitation of dates and times? It makes the information (in broad strokes at least) much more memorable, too. The fact that the Precambrian Eon is so much longer than the Phanerozoic Eon will be much more easily remembered if you take due note of this simple image, although the actual names and dates may require mnemonics. But using both means they support each other, making your learning that much stronger.

Let's make a further subdivision. The Phanerozoic Eon is divided into three eras:

- **Cenozoic Era**
- **Mesozoic Era**
- **Paleozoic Era**

The Cenozoic Era is the most recent, as indicated by its name, from the Greek *ceno*, meaning "new" (actually, Cenozoic can also be written Caenozoic or Cainozoic, and the Greek word is more traditionally rendered as *kainos*).

Mesozoic is derived from the Greek *mesos*, meaning middle, and is more easily remembered because a number of terms use the meso- prefix to indicate "middle", as in Meso-America.

Similarly, the paleo- prefix has also been used frequently to indicate that something is ancient (from the Greek *palaios*, meaning ancient).

In other words, the Phanerozoic Eon is divided very simply into an oldest period (Paleozoic — ancient life), a middle period (Mesozoic — middle life), and a new period (Cenozoic — new life).

What divides them? As with the division between the Precambrian Eon and the Phanerozoic Eon, which was marked by the Cambrian Explosion, the divisions within the Phanerozoic Eon are also caused by dramatic events. In this case, rather than an explosion of life, the divisions are marked by massive extinctions.

The most famous one is the so-called **K-T Boundary** event, 65 m.y. ago, and the reason why it is famous is not because of its size (although it was still impressive, wiping out some 85% of living species) but because of its most prominent victims: the dinosaurs. (The label "K-T" stands for Cretaceous-Tertiary.)

The other defining event was the Permian extinction, 250 m.y. ago. Although less well-known than the event that wiped out the dinosaurs, this was in fact the largest mass extinction Earth has ever seen, killing perhaps as many as 96% of all marine species, and more than 75% of vertebrate families on land.

So we have three defining events and three key dates:

- **Cambrian Explosion** 542 m.y. ago
- **Permian Extinction** 250 m.y. ago
- **K-T Extinction** 65 m.y. ago

These three dates are in fact relatively easy numbers to

remember; you may well find it takes little effort to impress them firmly into your mental database.

Let's consider these numbers and events in a little more detail. First of all, just to put these numbers in context, we should think about the fact that the Precambrian Eon spans the first *four billion* years of Earth's existence. The Phanerozoic Eon is only an eighth of that:

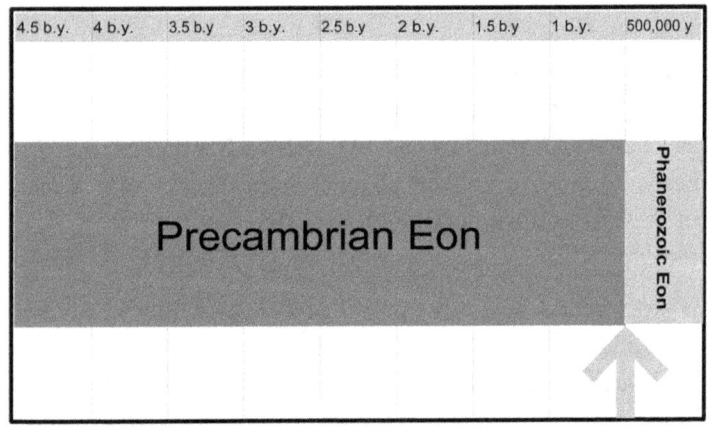

Cambrian Explosion
542 m.y.

As we travel forward in time, closer to the present day, the traces of events become more and more easy to read. Thus, we start by lumping together the entire first four billion years and focusing only on the most recent 542 million years. Now we refine our gaze a little further: a threefold division.

But our principle still applies: the Paleozoic Era covers more than half the length of the Phanerozoic Eon (292 m.y. out of 542 m.y.); the Mesozoic Era is a little over half the length of the Paleozoic Era (185 m.y.), and three times the length of the most recent era, the Cenozoic Era.

Visuals always help

Here it is expressed visually:

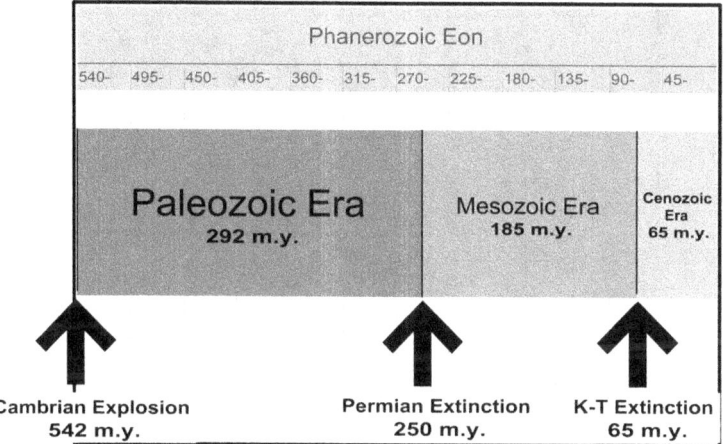

Now it's not necessary to learn all these numbers — if you know the anchor dates (**542 m.y.**, **250 m.y.**, and **65 m.y.**) you can work out everything else — but once again it's helpful to get a feel for the lengths of time we're dealing with, and in particular the *relative* lengths. Knowing that the Mesozoic Era is nearly three times the length of the Cenozoic Era, and that the Paleozoic Era covers more than half of the Phanerozoic Eon, keeps you on track.

On the next page you can see, in a neat box, the information you *must* have firmly in your head before going onto the next stage.

If you're having trouble remembering this information, here are a few more mnemonics to help you. But remember that mnemonics are best kept to a minimum — if you can remember without the help, do.

For these mnemonic images for the Eras, I have provided two

Eon	Era	Event Marker
Phanerozoic Eon 542 mya—Present	Cenozoic Era 65 mya—present	K-T Extinction 65 mya
	Mesozoic Era 250 mya—65 mya	Permian Extinction 250 mya
	Paleozoic Era 542 mya—250 mya	Cambrian Explosion 542 mya
Precambrian Eon 4565 mya—542 mya		

or three versions for each, with increasing amounts of information (I've only provided the middle variant for the first era, since it's such a simple difference). You only want the information you need help with; don't try to memorize a more complicated phrase or image, if a simpler one is all you need.

Notice in some variants that a red cross is used to indicate an error — our signifier for Era. The most complete picture includes a phrase containing the encoded date. You can, if you desire, easily create phrases for the simpler pictures, if that's all you require. For example, "**Palace zoo**", "**A zoo at the palace is an error**". If you want to incorporate a reminder of "Era" into the complete verbal phrase, you could expand the one given to something like: **In a ghastly error, a shell is going to hit the palace zoo.**

Glossary

boundary — límite
approximately — aproximadamente
dry — seco
in broad strokes — en líneas generales
due — adecuado
further — adicional
rendered — traducido
derived — derivado
ancient — antiguo
massive — masivo
extinctions — extinciones
impressive — impresionante
wiping out — aniquilando
victims — víctimas
mass extinction — extinción masiva
impress — impresionar
lumping together — concentrándose juntos
traces — huellas
refine — refinar, perfeccionar
on track — en camino de
neat — ordenado

Let's narrow our focus again, looking at one era at a time.

Paleozoic Era

Paleozoic Era 542—250 mya
  Permian Period 300—250 mya
  Carboniferous Period 360—300 mya
  Devonian Period 416—360 mya
  Silurian Period 444—416 mya
  Ordovician Period 488—444 mya
  Cambrian Period 542—488 mya

Cambrian Period

We've covered the Cambrian Period (named after Cambria, the Roman name for Wales; defined by the Cambrian Explosion). Earlier, I showed a mnemonic picture for the Cambrian Explosion. Here's a slightly different one, a very basic one, for the Period itself. See the black spot? That's a period (full stop), indicating that this is a Period, rather than an Era or an Eon. Look out for it in the next pictures.

Cambrian Period

The verbal equivalent of that period is "stop". So, for example,

Case Study   207

in the more complete mnemonic picture, we have the verbal phrase: **Campers learn when to stop**. "Learn", of course, is the encoded starting date, which you'll recall from the mnemonic for Cambrian Explosion. You don't need both. This one is provided as an example, but the one for Cambrian Explosion provides more information, and contains within it all you need to remember this (as long as you remember that the Cambrian Explosion marks the beginning of the Cambrian Period). You could combine both aspects into a verbal mnemonic such as: **The Welsh learn how to stop trilobites exploding in camp**. Or, if you don't feel the need to include the Welsh aspect, simply: **Learn how to stop trilobites exploding in camp.**

Cambrian Period

Ordovician Period

The next period is the Ordovician, which is named after a Celtic tribe called the Ordovices. This is probably not a lot of help to most of us. "Order" and "vice" are obvious similar words, but before we can use these we need to know what distinguishes the Ordovician Period.

Earth was a very different world then. Most of it was covered with water: north of the equator was almost entirely ocean; in the south, almost all the land was gathered into the supercontinent **Gondwana**, which spent the period drifting toward the South Pole, where much of it was submerged underwater.

So, a water planet; its life was marine. **Trilobites** were the dominant animals. The Ordovician Period ended with another mass extinction, in which 60% of all marine invertebrate genera and 25% of all families became extinct. Perhaps our slogan (mnemonic) for the Ordovician could be: **Order the vicious trilobites!**

Ordovician Period

The Ordovician began 488 million years ago (it's not necessary to learn two dates for each period if you know the order of periods — information more fundamental than exact dates, after all). If you're using a coding mnemonic, 488 translates as r-f-f — ruff off, perhaps (remember that doubled consonants generally count as one). **Order the vicious trilobite to take its ruff off!** Or, if you want to incorporate the signifier for "Period": **Order the vicious trilobite to stop taking its ruff off**. Notice the multiple black spots used — while one is sufficient, it is easily overlooked. The use of many makes it more obvious.

Silurian Period

The Silurian Period is likewise named after a Celtic Tribe, the Silures. "Silver" might be an appropriate keyword. What distinguishes the Silurian Period? The climate became more settled in the Silurian; large glacial formations melted, leading to a substantial rise in sea level. But first and foremost, the Silurian was a time for fish to dominate. Jawless fish spread widely; the first known freshwater fish appeared, as did the first fish with jaws. Additionally, we find our first good evidence of life on land.

The Silurian Period is dated from 444 million years (there's an easy number!). 444 = r-r-r. **The silver fish is a roarer?** Or **Stop! The silver fish is a roarer.**

Silurian Period

Devonian Period

The Devonian Period is named after the English county Devonshire, where the first Devonian rocks were found. The first land-living vertebrates appeared during the Devonian, and the first terrestrial arthropods.

"Devotion" makes a great keyword for Devonian, but is not so great for visualizing. You could use a church or prayer book to symbolize it, but it's usually better to go with something that is less ambiguous, even if it's not as perfect a sound-match.

There's another consideration as well, though it's premature to mention it. Later, we'll look at how to remember the order of periods. It turns out that "devil" is easier to work into this than "devotion".

The Devonian began 416 million years ago. 416 = r-t/d-ch/sh. **The devil offers the spider a radish?** To which you could add: **stop him**.

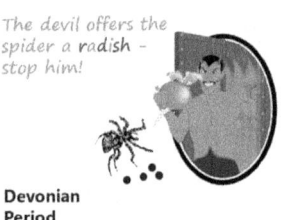
Devonian Period

Carboniferous Period

The Carboniferous Period is the Coal Age (carboniferous means coal-bearing). Apart from providing perfect conditions for making coal, the period is distinguished by the birth of the amniote egg. The amniote egg, by preventing the drying out of the embryo, allowed the first ancestors of reptiles, birds, and mammals to reproduce on land.

The Carboniferous Period began 360 million years ago. 360=m-sh-s. **There's a carbonated egg in my shoes? Stop the carbonated egg in my shoes?**

It should be noted that in the

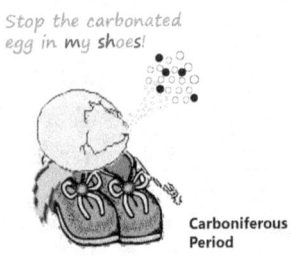
Carboniferous Period

United States they use two other periods in place of the Carboniferous Period: the **Pennsylvanian Period** (325-286 mya) and the **Mississippian Period** (360-325 mya).

Permian Period

And finally, we have the Permian Period, named after the Russian province of Perm, where rocks of this time were first studied. "Permanent" might be an appropriate keyword.

The Permian Period is of course best known for ending in the greatest of Earth's mass extinctions. It was also a time when earth's geography was dominated by one huge land mass (**Pangea**), which stretched from the northern to the southern pole, and one huge ocean (**Panthalassa**). The Permian Period began 300 million years ago. 300=m-s-s. **Permanent messes pong**? Pong is good because it recalls Pangea, as well as having an association with mess. However, it's a little harder to work in "Stop", if you want to do that. **Stop permanent messes stinking** rolls off the tongue better, but you could go with **Stop the pong from permanent messes**.

Summary

Time for a summary diagram — notice the sound bites associated with each period. These are gross simplifications, but that's what we're doing right now — we're simplifying, to set a framework in place. Don't be afraid of simplification (as long as you always remain aware that it is a simplification!). These sound bites tell a simple story. The story would be more consistent if I'd put "Birth of the amniotic egg" beside Carboniferous, but the period is associated so strongly with coal that I felt I couldn't ignore that.

Case Study

		Permian Period 300—250 mya	Everything dies
Paleozoic Era	300 mya		
	350 mya	Carboniferous Period 360—300 mya	Coal is laid down
	400 mya	Devonian Period 416—360 mya	Animals crawl onto land
	450 mya	Silurian Period 444—416 mya	Fish take over
	500 mya	Ordovician Period 488—444 mya	Trilobites rule the world
	550 mya	Cambrian Period 542—488 mya	**Life explodes**

The graphic also gives us some idea of the relative length of periods — you can see that, unlike our earlier divisions, these segments are fairly even. The Silurian Period is distinctly shorter than the rest, but not hugely so. Here are the lengths:

- Permian 50 my
- Carboniferous 40 my
- Devonian 56 my
- Silurian 28 my
- Ordovician 44 my
- Cambrian 54 my

So you can come away with a general understanding that the Paleozoic Era — lasting nearly 300 million years — is divided into six periods that are roughly equal in length, around 50 million years, with the Silurian Period being decidedly shorter than the rest. The Era begins with an explosion of life, sees trilobites being succeeded by fish, sees the land colonized, coal laid down, and then, sees almost all its marine life and the majority of its terrestrial life extinguished.

You may like to provide yourself with a first-letter acrostic for these six periods — far easier to remember than the lengthy acrostic you need for all the periods of the Phanerozoic Eon.

Campers **Ord**er Silver **Dev**ils **Carbon**ated **Perm**anently

(**COSDCP**, if you think the initials are sufficiently memorable.)

But it would probably help to tag these with a reminder that they all belong in the Paleozoic Era, so better to say:

In the **Pal**ace, **Cam**pers **Ord**er Silver **Dev**ils **Carbon**ated **Perm**anently.

We're ready now to consider our earlier question about whether you need a coding mnemonic to remember the important dates. It's not a question that can be answered with a simple yes or no; it is a matter for the individual. Some people are comfortable with numbers and find them easy to remember (I imagine there is a connection between these two attributes!). If you are one of those, I do believe the dates involved are in fact quite easy to master:

- 542; 250; 65
- 542; 488; 444; 416; 360; 300; 250

If you don't find the dates easy to remember, grouped in this way, then I suggest you use a coding mnemonic.

Glossary

full stop — punto

tribe — tribu

drifting — deriva

submerged — sumergido

underwater — submarino

settled — estable

jawless fish — pescado sin mandíbula

jaw — mandíbula

coal — carbón

drying out — el secado

embryo — embrión

mammals — mamíferos

pong — hedor

rolls off the tongue — sale de la lengua

sound bite — frase incisiva

gross — brutas

framework — marco

decidedly — decididamente

succeeded by — seguido por

majority — mayoría

extinguished — extinguido

comfortable — cómodo

Time to review

There are two aspects to review. There is, obviously, your practice / testing of the content, the information you're trying to learn. But if you're using mnemonics, the first type of practice you need to do is of the mnemonics themselves. Only when you've mastered these should you go on to testing your learning of the material.

So how do you practice mnemonics?

As with any practice, the first and most critical point is to make sure you're practicing the right thing.

To do that, you need to think about the circumstances in which you'll be trying to recall the information. For example, you might want to recall:

1. the dates for the Devonian Period

2. which Period follows the Carboniferous

3. which Period began 488 mya

4. when fish rose.

For the first, if using mnemonics, you'd need to remember the keyword for Devonian (devil), remember the verbal phrase associated with that (whether via the image or not), and then decode the word carrying the date information. Since that gives you the starting date only, you then need to remember which Period follows the Devonian, and then go through the same process for that (its starting date being, of course, the end-date for the Devonian).

In the third example, you'd need to encode 488 and recognize the encoded word used, evoking the phrase incorporating that, then recognizing the keyword, and recalling from that, the name of the Period.

For the last, you'd need to remember the phrase or image that highlights fish, working backward as it were, so that fish evokes silver, which in turn evokes Silurian.

All of this sounds more complicated than it really is. Here are the specific links:

1. Devonian — devil — The devil offers the spider a radish — radish = 416; Campers Order Silver Devils Carbonated — Stop the carbonated egg in my shoes — my shoes = 360

2. Carboniferous — Carbonated — Campers Order Silver Devils Carbonated Permanently — Permian

3. 488 = r-t/d-sh/ch — radish — The devil offers the spider a radish — devil — Devonian

4. fish — The silver fish is a roarer — silver — Silurian

What all this means is that you want

- the name of the Period to reliably evoke the keyword
- the keyword to reliably evoke the image &/or the phrase
- aspects of the image/phrase to reliably evoke the keyword
- the keyword to evoke the acrostic giving the order.

And if you are using the coding mnemonic, you do of course want to be able to reliably and quickly decode.

So, for example, if you were using flashcards, some might have the name of a Period on one side, and the keyword on the other (e.g., Ordovician Period / Order), and you would practice in both directions. Other cards might have a coded word on one side and the date on the other (e.g., Roarer / 444).

Do remember, though, that you also need to practice recalling the image. Thus, when responding to a keyword, you don't simply want to recall the word or phrase associated with it — you also want to visualize the mnemonic image. You also want that image to reliably evoke the keyword.

To assist you with your practice for that, here are the mnemonic images we've covered, without any identifying words or phrases:

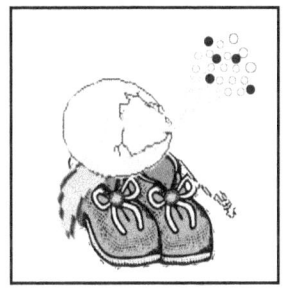

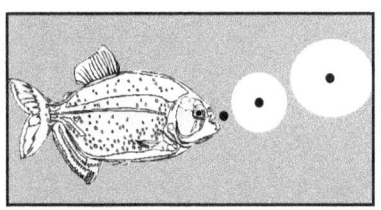

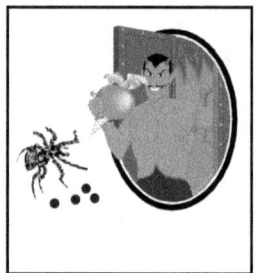

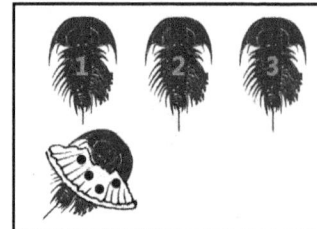

Remember also that you only want to use mnemonics when you need the help. Don't bother using them for information you can easily remember.

Once you're confident of any mnemonics you're using, you're ready to test your hold on the content. Here's a short multi-choice review to test that.

Review 12.1

1. What event subdivides the most recent eon from earlier eons?

    a) K-T Extinction

    b) Cambrian Explosion

    c) Rise of mammals

    d) Permian Extinction

2. What is the date for this event?

    a) 65 mya

    b) 249 mya

    c) 250 mya

    d) 542 mya

3. What is the name of the most recent eon?

    a) Phanerozoic

    b) Cenozoic

    c) Holocene

    d) Mesozoic

4. Into how many parts is this eon subdivided?

    a) 4

    b) 3

    c) 6

    d) 2

5. What are the names for these parts?

    a) Mesozoic Era

    b) Paleozoic Era

    c) Proterozoic Eon

    d) Cenozoic Era

6. What events divide these parts?

    a) Rise of mammals

    b) Permian Extinction

    c) Rise of dinosaurs

    d) K-T Extinction

7. What are the dates for these events?

    a) 250 mya

    b) 542 mya

    c) 56 mya

    d) 65 mya

8. Into how many parts is the oldest subdivision of the most recent eon divided?

    a) 6

    b) 4

    c) 2

    d) 3

9. Name these parts in order from oldest to most recent: Cambrian, Carboniferous, Devonian, Ordovician, Permian, Silurian

10. Match the starting dates with the correct Period.

Devonian	444 mya
Cambrian	300 mya
Ordovician	416 mya
Permian	542 mya
Carboniferous	488 mya
Silurian	360 mya

If you answered all these correctly and without hesitation, you are ready to move on. If you didn't, I suggest you go over the material, using the mnemonics, or spending more time on them, for each piece of information not thoroughly absorbed.

The Mesozoic Era

> **Mesozoic Era 250—65 mya**
> Cretaceous Period 145—65 mya
> Jurassic Period 200—145 mya
> Triassic Period 250—200 mya

Triassic Period

The Mesozoic Era is, like the Phanerozoic Eon, divided into

three parts. The oldest is the Triassic Period, named for the threefold division of rocks of this age in Germany. It began, of course, 250 million years ago, and may be principally thought of as a time of transition, as the world recovered from the massive Permian extinction. Dinosaurs began their rise to domination, and the super-continent Pangea began to drift apart. 250=n-l-s. **Your triangle's drifting — try nails.**

Notice the additional details in this mnemonic picture, depicting an ass hiding behind a triangle. This ass (rear end) of an ass (donkey) is the signal for the 'assic' ending, and it'll be used again in the next image.

Jurassic Period

The Triassic Period was succeeded by the Jurassic Period, which probably needs no help to be remembered since the famous Spielberg movie *Jurassic Park*!

It is named after the Jura Mountains between France and Switzerland, where rocks of this age were first studied. As the movie connection suggests, the Jurassic period could be subtitled the "Age of Dinosaurs". It also saw the birth of birds, and teleost fish. Pangea broke apart into Gondwanaland (in the south) and **Laurasia** (in the north).

The Jurassic Period began 200 million years ago. 200=n-s-s. **Dinosaur noses!**

Or, if you need a reminder of the name: **Spot the dinosaur noses on the jury!**

Cretaceous Period

The final and most recent period of the Mesozoic Era is the Cretaceous Period, named from the Latin word for chalk (creta) and first applied to extensive deposits of this age that form the famous white cliffs along the English Channel. This period is also sometimes named the Age of Dinosaurs; it did of course end with their extinction (the K-T Extinction). Flowering plants arose in the Cretaceous, and colonial insects. North America and Eurasia separated, and South America and Africa separated. The Cretaceous began 145 million years ago. 145=t-r-l. **Trial of the dinosaur's crate?**

trial of the dinosaur's crate?

Or, if you think it's more useful, you could instead visualize a crate of chalks.

Time for another summary diagram.

Mesozoic Era			
	100 mya	Cretaceous Period 145 mya—65 mya	Dinosaurs die
	150 mya		
	200 mya	Jurassic Period 200 mya—145 mya	Dinosaurs rule the world
	250 mya	Triassic Period 250 mya—200 mya	Life recovers

Notice that the Triassic and Jurassic Periods lasted a similar length of time, about the same length of time as the Paleozoic periods, while the Cretaceous Period was distinctly longer.

Okay, let's recap. The Phanerozoic Eon ("visible life") has so far lasted 542 million years. The Paleozoic Era covered the first 292 million years of that. From 250 mya (Permian Extinction) to 65 mya (K-T Extinction), a period of 185 million years, was the Mesozoic Era — a time which can fairly be described as the Age of the Dinosaurs. The Mesozoic Era is divided into three parts: the Triassic, lasting 50 million years; the Jurassic, lasting 55 million years; the Cretaceous, lasting 80 million years.

If you need some help remembering the order of the Periods, you could go with: **Try the Jury in a Crate!** or, if you want it tagged with a cue to the Era, perhaps: **The Message said to Try the Jury in a Crate.**

When you're confident you've mastered any mnemonics you need, see how you go on the content review. (Mnemonic images without identifying words are provided for your practice at the end of the chapter.)

Remember, if you can't answer these questions without hesitation, you need to review until it's all firmly in your head.

Review 12.2

1. What's the shortest period in the Paleozoic Era?

   a) Devonian

   b) Silurian

   c) Cretaceous

   d) Precambrian

2. When did the largest mass extinction in Earth's history occur?

   a) 250 mya

   b) 145 mya

   c) 542 mya

   d) 65 mya

3. Dinosaurs arose during which of these periods?

   a) Cambrian

   b) Jurassic

   c) Cretaceous

   d) Triassic

4. How long was the Precambrian?

   a) 50 million years

   b) 80 million years

   c) 5 billion years

   d) 4 billion years

5. Of all the Paleozoic and Mesozoic Periods, which is the longest?

   a) Cambrian

   b) Cretaceous

   c) Carboniferous

   d) Devonian

6. What event marks the end of the Paleozoic Era?

   a) Cambrian Explosion

   b) Appearance of the amniotic egg

   c) K-T Extinction

   d) Permian Extinction

7. How long ago did dinosaurs disappear from the earth?

   a) 145 mya

   b) 65 mya

   c) 250 mya

   d) 200 mya

8. South America and Africa separated during which period?

   a) Permian

   b) Jurassic

   c) Ordovician

   d) Cretaceous

9. Trilobites dominated earth during which period?

   a) Ordovician

   b) Carboniferous

   c) Jurassic

   d) Permian

10. The present eon began when?

   a) 65 mya

   b) 250 mya

   c) 542 mya

   d) 488 mya

Glossary

evoking — evocando

highlights — destaca

bother — te molestes

hesitation — vacilación

thoroughly — completamente

absorbed — absorbido

transition — transición

recovered — recuperado

recap — recapitulemos

disappear — desaparecer

Cenozoic Era

> **Cenozoic Era 65 mya—present**
> **Neogene Period 23 mya—present**
> Holocene Epoch 8000 ya—present
> Pleistocene Epoch 1.8 mya—8000ya
> Pliocene Epoch 5.3—1.8 mya
> Miocene Epoch 23—5.3 mya
> **Paleogene Period 65—23 mya**
> Oligocene Epoch 34—23 mya
> Eocene Epoch 56—34 mya
> Paleocene Epoch 65—56 mya

The modern era — the Cenozoic Era — began 65 million years ago. The extinction of the dinosaurs, so long dominant, allowed the rise of other species.

The Cenozoic Era was originally divided into Primary, Secondary, Tertiary, and Quaternary. The first two names were then dropped, but, confusingly, the terms **Tertiary** and **Quaternary** were kept as period designations. You will still see these names used, but the official subdivisions of the Cenozoic Era are now known as the **Paleogene** and **Neogene** Periods ("old" and "new" respectively; the 'gene' ending derives from a Greek word meaning 'of a specified kind' — again, a singularly unhelpful term).

We're now closing in on modern times, so of course we become more detailed. The two periods of the Cenozoic Era (a mere 65 million years in total — hardly longer than any single period of the Paleozoic or Mesozoic Era) are subdivided into epochs. The Paleogene Period encompasses the **Paleocene Epoch** (yes, you need to pay particular attention to that little g/c difference!), the **Eocene Epoch**, and the **Oligocene Epoch**. The Neogene period encompasses the **Miocene Epoch**, the **Pliocene Epoch**, the **Pleistocene Epoch**, and the (current) **Holocene Epoch**.

We looked at these briefly earlier; now let's go into a bit more depth.

Paleocene Epoch

Here's a fuller mnemonic image for this epoch, providing visuals for the phrase: **Dinosaurs hide in the palace shell**. Palace is of course our keyword for Paleocene. Shell is code for the starting date: 65 mya. Notice, too, the blue

Paleocene Epoch

pocket in this picture: this pocket is the sign for an Epoch. The mnemonic commemorates the fact that the K-T Extinction event marks the start of this epoch.

The Paleocene Epoch ends with a less dramatic event, but one of great significance nevertheless: the **Paleocene-Eocene Thermal Maximum** (also known as Eocene thermal maximum 1, ETM1). This event, generally dated to around 55.5 mya, is thought to have lasted no more than

Paleocene-Eocene Thermal Maximum

20,000 years. During it, a massive release of carbon increased global temperatures by 5-8°C.

Eocene Epoch

The oldest known fossils of most of the modern orders of mammals appear in the Early Eocene. This time period was also the hottest period of the Cenozoic Era; the world was basically free of ice (and so had a lot of rain! Eeyore's tears might remind us of that). Antarctica and Australia separated near the end of the Eocene Epoch, creating changes in the ocean's circulation patterns that cooled the world and resulted in more climate variability. These factors are thought to have driven the increase in mammalian body size.

Here's the full mnemonic image for this epoch, providing visuals for the phrase: **Eeyore's leash parts the continents**. Leash is code for the starting date: 56 mya.

Eocene Epoch

Oligocene Epoch

Many grasses appeared in the Oligocene, and so did early horses and trunked elephants. Mammals began to dominate the land (except in Australia). Herbivores continued to grow in size with the increase in grassland. The first primates appeared.

More olives for elephants! "More", of course, signifies the starting date: 34 mya.

Oligocene Epoch

Miocene Epoch

The Miocene was warmer than the Oligocene; grasslands spread, and kelp forests appeared. Africa joined up with Asia, enabling animals to migrate between. The epoch began in 23 mya, so: **Name who spilt milk on the grass**.

Pliocene Epoch

The Pliocene was again cooler. Grasslands and savannas spread widely. Hominids arose. The Panama landbridge joined North and South America, allowing animals to migrate between. At the poles, ice accumulated, eventually killing most of the animals that lived there. Glaciers advanced in other parts of the world. **Lo! Some pliable hominids arose.**

"Lo! Some" signifies the starting date: 5.3 mya. Remember that *s* at the beginning of a word indicates a decimal point.

Pleistocene Epoch

The Pleistocene is known as the Great Ice Age; these are the ice ages we think of when we talk of "the ice age" — the time of woolly mammoth, the saber-tooth, the mastodon.

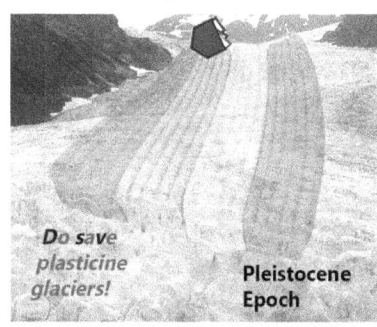

This is the epoch when Homo sapiens rose. By the end of this epoch, humans had spread over much of the world, and the giant mammals were dead. **Do save plasticine glaciers!** (Do save = 1.8 mya)

Holocene Epoch

Modern man faces ace in a hole! "Faces ace" encodes 8000 — a bit tricky getting those 3 zeroes in a row!

Time for another summary diagram.

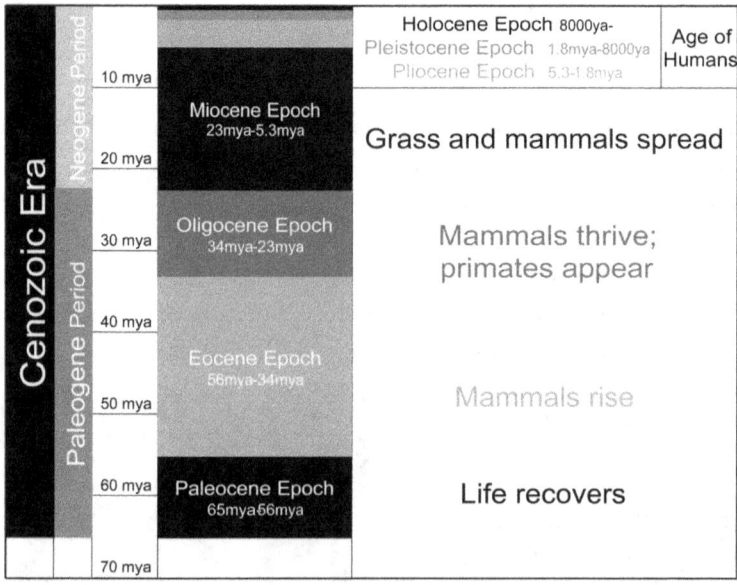

Notice the Paleogene Period is of comparable length to periods in the Paleozoic and Mesozoic Eras, and roughly twice the length of the Neogene Period. Notice too, the respective lengths of the epochs. The Eocene is the biggest, although the Miocene isn't far behind — but even these are less than half the average length of a period. The Paleocene and Oligocene Epochs are only around 10 million years each, and the most recent epochs are so short I can't fit their names on the scale.

The dates for these time periods are not as easy to remember as those of earlier times; nevertheless, grouping makes a huge difference.

Thus, you have two dates to remember for the basic subdivision of Periods: 65 mya for the Paleogene Period (which you already know because it's one of your "anchor" dates: the K-T Boundary event), and 23 mya for the Neogene. You need two dates to mark the subdivisions of the Paleogene (56 and 34 — notice the pattern here: 56, 34, 23 — all sequential numbers). You need three dates for the Neogene: 5.3; 1.8; 8000. Of all the dates we've had to remember, these are the hardest. But three hard dates out of 19 is not so bad!

A possible acrostic to help you remember the order of the Epochs is: **For a Cent, Pale Eeyore tossed Olive Milk down the Pliable Plasticine Hole**.

Time to see how well you've mastered all this material (remember to practice your chosen mnemonics first! — note the remaining practice mnemonic images are displayed after the final review, if you're wanting them.

Review 12.3

Select the correct time-period for the event:

1. North America separates from Eurasia
   a) Miocene Epoch
   b) Pliocene Epoch
   c) Cretaceous Period
   d) Cambrian Period

2. Trilobites appear
   a) Silurian Period
   b) Cambrian Period
   c) Ordovician Period
   d) Jurassic Period

3. Fish dominate the seas
   a) Silurian Period
   b) Devonian Period
   c) Triassic Period
   d) Miocene Epoch

4. Almost all sea animals extinguished, and most land vertebrates
   a) K-T Extinction
   b) Permian Extinction

c) Devonian Period

   d) Cambrian Explosion

5. Africa and Asia joined up

   a) Pliocene Epoch

   b) Cretaceous Period

   c) Jurassic Period

   d) Miocene Epoch

6. Homo sapiens appears

   a) Pleistocene Epoch

   b) Oligocene Epoch

   c) Pliocene Epoch

   d) Jurassic Period

7. Age of Dinosaurs

   a) Triassic Period

   b) Pleistocene Period

   c) Paleozoic Era

   d) Mesozoic Era

8. The Americas joined up

   a) Cretaceous Period

   b) Miocene Epoch

c) Pliocene Epoch

   d) Permian Period

9. The first land vertebrates and arthropods appear

   a) Ordovician Period

   b) Devonian Period

   c) Triassic Period

   d) Eocene Period

10. Primates appear

    a) Oligocene Epoch

    b) Pleistocene Epoch

    c) Pliocene Epoch

    d) Permian Period

Review 12.4

Match the event with the starting date.

1. Paleogene Period

   a) 56 mya

   b) 145 mya

   c) 5.3 mya

   d) 65 mya

2. 23 mya
   a) Miocene Epoch
   b) Triassic Period
   c) Oligocene Epoch
   d) Pliocene Epoch

3. Cretaceous Period
   a) 360 mya
   b) 65 mya
   c) 542 mya
   d) 145 mya

4. 56 mya
   a) Paleocene Epoch
   b) Carboniferous Period
   c) Eocene Epoch
   d) Cretaceous Period

5. Cenozoic Era
   a) 360 mya
   b) 65 mya
   c) 542 mya
   d) 145 mya

6. Phanerozoic Eon
   a) 542 mya
   b) 300 mya
   c) 2500 mya
   d) 65 mya

7. 200 mya
   a) Permian Period
   b) Jurassic Period
   c) Triassic Period
   d) Paleogene Period

8. Permian Period
   a) 56 mya
   b) 250 mya
   c) 34 mya
   d) 300 mya

9. Silurian Period
   a) 444 mya
   b) 416 mya
   c) 542 mya
   d) 145 mya

10. 488 mya
   a) Triassic Period
   b) Ordovician Period
   c) Cambrian Period
   b) Devonian Period

Assuming you completed these quickly and easily, you're now ready to leave the Phanerozoic Eon and take a look at the Precambrian.

Glossary

singularly unhelpful — realmente muy inútil

release — emisión

circulation patterns — patrones de circulación

leash — correa

trunked elephants — elefantes troncos

grasslands — pastizales

kelp forest — bosque de algas marinas

landbridge — puente de tierra

woolly mammoth — mamut lanudo

saber-tooth — tigre dientes de sable

mastodon — mastodonte

Precambrian

> **Precambrian 4560—542 mya**
> **Proterozoic Eon 2500—542 mya**
> **Archean Eon 3800—2500 mya**
> **Hadean Eon 4560—3800 mya**

It's worth noting that recently another term is sometimes heard in place of "Precambrian", and that is "**Cryptozoic**", meaning "hidden life". This fits in nicely with the Phanerozoic ("visible life"), and has come about in response to recent discoveries of complex life predating the Cambrian Explosion.

You may notice I haven't called this the "Precambrian Eon". You will see it sometimes called this, but technically the Precambrian is an informal title, encompassing two (or more) eons. The International Commission on Stratigraphy (whose scale is the one I'm following) recognizes only two. However other sources include a third, and I rather like the symmetry of having three! (remembering the Phanerozoic Eon is divided into three).

Moreover, and probably somewhat more to the point, the **Hadean** Eon encompasses what has been termed "pre-geologic" time — the period when Earth was formed. It makes sense then to consider this time period as separate from later periods. You may prefer, however, to keep faith with the International Commission, and so I note that they subsume the Hadean as a sub-period within the **Eoarchan Era** (~ 3600 mya), the first era of the **Archean Eon**.

Hadean Eon

What we know of Hadean events is largely derived from

studies of lunar geology (during the Hadean, the Moon was a mere 16,000 km from the Earth, compared to the 384,000 km between us today). We do however have some record of Hadean times on Earth — the most ancient rocks found on Earth to date were found in Australia, and are 4.2 billion years old.

Our beginning date is perhaps spuriously accurate: 4560 million years ago (4 ½ billion, essentially). The International Commission prefers to leave the beginning open, but the figure does seem generally accepted. It's often rounded as 4.6 billion, which is no doubt perfectly adequate.

All of this is by way of explaining some inconsistencies you might have found in names and dates, but it's not something you need to memorize. What we need to know is that Earth's timeline begins with the Hadean Eon, and so its starting date is the starting date of our planet also.

The name "Hadean" is derived from Hades, the Greek mythological hell — a fitting name, considering the volcanic events of this time.

While Hades is a great mnemonic aid for this Eon (as long as you're familiar with it), it's not so easy to portray. Accordingly, our visual mnemonic uses another keyword: aid. The starting date is encoded as "relishes" in the mnemonic phrase: **relishes giving aid**.

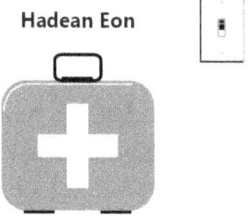

Hadean Eon

relishes giving aid

If you are familiar with Hades, you could include it in this phrase to reinforce the idea of volcanic birth: **relishes giving aid in Hades** (or simply: **relishes Hades**). A somewhat busier mnemonic picture could incorporate this idea, enabling us to tell a fuller story, one where the aid is needed because of the volcanic activity.

Case Study

Note the light switch in the picture — this is our signifier for Eon ("on"). Imagine someone turning the switch on, thus starting the volcanoes all over Earth. It's hellish, but look, here go the ambulance men rushing to give aid (I'm picturing World War I medics making their way across the devastated landscape).

Hadean Eon

You see how you can bring your own experiences and perceptions to mnemonic phrases and pictures, to make a fuller story in your mind. The more of an emotional reaction you have, the better your memory of it will be. Just remember to hold on to your focuses — the words and images holding the key information. The story you tell yourself is supporting detail that will be remembered more easily, but you don't want it to overwrite the information you need!

Archean Eon

The Archean Eon covers the time when most continental landmasses (70%) were formed, although they were probably submerged at this time. The Earth is thought to have been all ocean. The landmasses formed at this time are known as "shields", and these form the cores of present-day continents.

The first microorganisms were anaerobic, using methane or hydrogen in their metabolism rather than oxygen. This is dramatically less effective than using oxygen, so it was just as well that organisms came along that could produce oxygen, setting the stage for later complex life. These organisms were still single-celled, however.

In other words, this was a very primitive time — hence the name. "Archean" means ancient or primitive (think of archaic).

"Archaic" might be a perfectly adequate keyword if all you want is a gentle reminder of the name. In the mnemonic picture, however, I've used "arch", it being much easier to visualize (or indeed, draw!). Having 2 zeroes in the starting date has its usual complication — we can't start a word with that s/soft c sound, because that would indicate a decimal point. Nor are we likely to come up with a single word that could incorporate that ending preceded by only the consonants m and f/v. Therefore, the second word must start with that f/v, leaving the m alone for the first word. I could have picked "my voices", but again, that would be harder to portray (if your mnemonic is purely verbal, you don't have that constraint). So, our mnemonic: **The arch holds my faces**.

I could have also gone with **The archer knows my faces**, which would have made more sense, but would have been a little harder to portray. Creating mnemonics, especially ones that incorporate encoded numbers, are often a matter of compromise. What will work best for you will always be an individual decision.

On the other hand, if you want to incorporate the idea of shields —the continents forming under the ocean — then archer fits in very nicely with this. So though it's a little wordier and harder to portray (though easy enough to mentally visualize), you could go with this fuller mnemonic: **Shield my faces from the archer**.

Proterozoic Eon

The Proterozoic Eon is the time when plate tectonics began to take control of determining the form of the Earth's crust. Continental drift was much faster then, with the Earth's magma so much hotter and closer to the surface.

Proterozoic means "early life", and it was during this time that single-celled organisms developed a nucleus, another critical step in the road to complex life. Near the end of this lengthy eon (two billion years!), multi-celled lifeforms finally appear.

Meaning-wise, "protean" makes a good keyword for this Eon, but because of our usual considerations (ability to visualize, and familiarity with the word), I've gone instead with the word "protect". This also fits in nicely with the 'zoo'. So, with "no losses" encoding the start date, here's our mnemonic phrase: **I protect the zoo — no losses**.

You could also bring in the idea of algae bringing life to the planet, by imagining an algal bloom spreading across the zoo. There's the guard, oblivious, thinking he's protecting the zoo by checking everyone's locked in, while the algae spreads insidiously across their feet.

If you want some help in remembering the order of these Eons, try: **Before Camping, Aid the Archer Protecting us**.

Summary

In this summary diagram for the Precambrian, notice the

length of the Proterozoic. If you follow the International Commission and subsume the Hadean with the Archean Eon, the Precambrian is essentially divided into two 2-billion-year periods. It also means you only need to remember two new numbers: 4560 (or 4600 if you want to simplify) and 2500.

Precambrian			
1 mya			
1.5 mya	Proterozoic Eon 2500 - 542 my	Green algae bring life	
2 mya			
2.5 mya			
3 mya	Archean Eon 3800 - 2500 my	Continents form under the ocean	
3.5 mya			
4 mya	Hadean Eon 4560 - 3800 my	Earth formed	
4.5 mya			

The big picture

Time to put it all together. Here's the full summary diagram — note that, while the width of the labels for the Paleozoic, Mesozoic and Cenozoic Eras, and the Precambrian, vary to indicate the relative lengths, it is of course impossible to fairly represent the timescale in one picture, given that the

Precambrian covers some 90% of the entire period!

Eon	Era	Sub-era	Period/Epoch	Description	
Phanerozoic Eon	Cenozoic Era 65 mya	Neogene	Holocene Epoch	Historic period	
			Pleistocene Epoch	Humans appear	Age of Humans
			Pliocene Epoch	Hominids appear	
		Paleogene	Miocene Epoch	Grass and mammals spread	
			Oligocene Epoch	Mammals thrive; primates appear	
			Eocene Epoch	Mammals rise	
			Paleocene Epoch	Life recovers	
	Mesozoic Era 250-65 mya		Cretaceous Period	Dinosaurs die	
			Jurassic Period	Dinosaurs rule the world	
			Triassic Period	Life recovers	
	Paleozoic Era 542-250 mya		Permian Period	Everything dies	
			Carboniferous Period	Coal is laid down	
			Devonian Period	Animals crawl onto land	
			Silurian Period	Fish take over	
			Ordovician Period	Trilobites rule the world	
			Cambrian Period	Life explodes	
Precambrian 4560-542 mya			Proterozoic Eon	Green algae bring life	
			Archean Eon	Continents form under the ocean	
			Hadean Eon	Earth formed	

Take careful note of the time periods used. For the Precambrian, our "base" periods are Eons; for the Paleozoic and Mesozoic Eras, they are Periods; for the Cenozoic Era, they are Epochs. This is more logical than it sounds — "eon" is used for the really dramatic changes; eons are divided into eras, often based on a change in which animals are dominant; eras are split into periods; periods into epochs. Because we become more interested in changes the closer we get to the modern-day, we get this effect that the most recent divisions are epochs, while the most distant divisions are eons.)

Of course there are further subdivisions for all these time periods, but these are the important ones, so this is your basic framework. These are what you should learn so well that you will never forget them, as well as the sound-bite story associated with them.

Do you have the key dates locked in, too?

The big picture: 4560; 542; 250; 65

Precambrian: 4560; 3800; 2500; 542 (or simply, 4600; 2500; 542)

Paleozoic: 542; 488; 444; 416; 360; 300; 250

Mesozoic: 250; 200; 145; 65

Cenozoic: 65; 56; 34; 23; 5.3; 1.8; 8000 (you have to remember all the others are in millions, but not this last)

(If you're having trouble remembering the order of Eon, Era, Period, Epoch, you could try the acronym **EEPE**, or the phrase **On is Wrong so Stop to Pack** (because while Pocket worked well as a visual, it's hard to get it to make sense here).)

Let's see how well you do on the final review. (Note that the remaining practice mnemonic images can be found at the end of the chapter.)

Review 12.5

1. What do the periods in the geologic time scale represent?

    a) times between extinction events

    b) a change in the dominant animals of the time

    c) the most dramatic changes in the history of the Earth

    d) a change in the plants and animals during that time

2. What do the terms Paleozoic, Mesozoic and Cenozoic mean?

    a) old new, less new, most new

    b) old zoo, middle zoo, modern zoo

    c) ancient life, middle life, recent life

    d) Stone Age, Middle Stone Age, Bronze Age

3. How is relative age different from the actual date of an event?

    a) Relative age only tells us the order in which events occurred, from the earliest to the most recent. Knowing the actual date of an event allows us to say exactly how old something is or how long ago it actually took place.

    b) Relative age tells us how long ago something happened relative to us today, and thus how old it is. The actual date is fixed, being counted from the beginning of the Earth.

4. What is the name of the current geological time-period?

   a) Phanerozoic Eon
   b) Neogene Period
   c) Cenozoic Era
   d) Holocene Epoch

5. Name the eras in order of length, with the longest first.

   a) Proterozoic, Archean, Hadean
   b) Paleozoic, Cenozoic, Mesozoic
   c) Paleozoic, Mesozoic, Cenozoic
   d) Precambrian, Paleozoic, Paleocene

6. What event separates the Phanerozoic Eon and the Precambrian, and how long ago did it occur?

   a) Cambrian Explosion, 250 mya
   b) Cambrian Explosion, 542 mya
   c) Permian Extinction, 250 mya
   d) K-T Boundary Event, 65 mya

7. What event separates the Paleozoic Era from the Mesozoic, and how long ago did it occur?

   a) Perrmian Extinction, 250 mya
   b) K-T Boundary Event, 65 mya

c) Cambrian Explosion, 250 mya

   d) Cambrian Explosion, 542 mya

8. What event separates the Mesozoic Era from the Cenozoic, and how long ago did it occur?

   a) Permian Extinction, 250 mya

   b) K-T Boundary Event, 65 mya

   c) Cambrian Explosion, 250 mya

   d) Cambrian Explosion, 542 mya

9. When was coal laid down?

   a) Cambrian Period

   b) Triassic Period

   c) Carboniferous Period

   d) Silurian Period

10. During what era and epoch did hominids rise?

    a) Paleocene Epoch, Cenozoic Era

    b) Pleistocene Epoch, Cenozoic Era

    c) Pleistocene Epoch, Mesozoic Era

    d) Pliocene Epoch, Cenozoic Era

I hope you've found this an instructive exercise, whether you simply read the material or fully participated in the learning process. It may have seemed somewhat overwhelming at times, but I hope you soldiered through it, if so.

The point I hope I've got across is that you need to think carefully about what you want or need to learn, and the amount of effort you want to put into it. Mnemonics aren't a silver bullet, but they are a valuable adjunct. They work best if you keep them to a minimum, using more meaningful strategies where you can, using both meaningful and mnemonic strategies to support each other.

I hope that participants also came to realize that, although the mnemonics may at first have seemed complicated or hard to create, you became more comfortable with them over time. It really is just a matter of practice.

This particular example may not have been to your taste, but now that you've seen the principles at work, you'll be able to apply them in all sorts of different situations.

Happy learning!

Glossary

symmetry — simetría

subsume — subsumen, incluyen

spuriously — engañosamente

relishes — disfruta

shields — escudos

setting the stage — preparando el escenario

plate tectonics — placas tectónicas

protean — proteico, mudable

fits in nicely — se adapta bien

algal bloom — floración de algas

oblivious — inconsciente de

insidiously — insidiosamente

soldiered through — tercamente empujado

silver bullet — bala de plata

Mnemonic images for practice

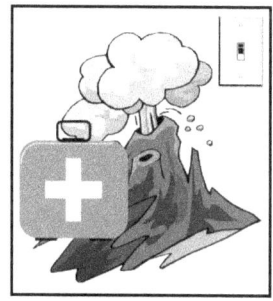

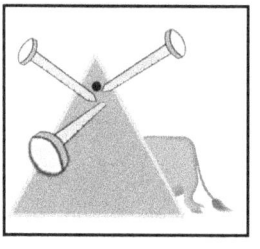

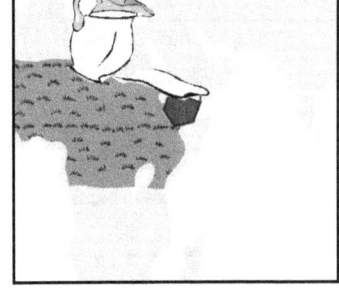

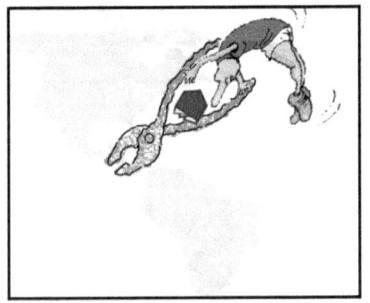

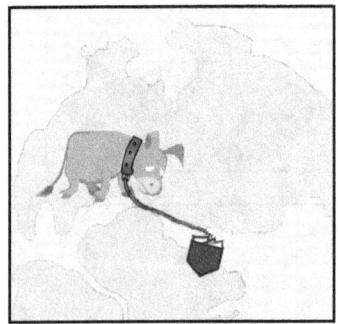

254   Mnemonics for Study

Answers

Review 1.1

1. a, b, e
2. a, b, e
3. c
4. c, d
5. a, c
6. F
7. e
8. c

Review 2.1

1. b, c, e
2. b, e
3. F
4. c, d
5. c, d, e

Review 3.1

1. a, c, d
2. b, d, e
3. a, b

Review 4.1

1. d
2. c
3. d
4. a, c, d, e, f
5. a, b, e, f
6. e
7. c, d
8. b

Review 5.1

1. b, c, d
2. a, c, d, e
3. b, d

Review 6.1

1. a, c, d
2. b
3. c
4. a, b, d, e

Review 7.1

1. b, d
2. c, d, e
3. a, d

4. b, c

Review 8.1

1. d
2. tree—3, bun—1, hen—10, sticks—6, shoe—2, wine—9, gate—8, door—4, hive—5
3. d
4. b
5. c, d

Review 9.1

1. b
2. b, d

Review 10.1

1. b, d
2. c
3. d
4. a
5. b
6. b
7. a
8. b
9. b
10. a
11. d

12. a
13. b
14. d
15. a, b, c, d

Review 11.1

1. c, e
2. a, b, c, d, e
3. F
4. T
5. a, b, c, d, e
6. c, d

Review 12.1

1. The Cambrian Explosion
2. 542 mya
3. Phanerozoic Eon
4. 3
5. Paleozoic Era; Mesozoic Era; Cenozoic Era
6. Mass extinctions: the Permian Extinction and the K-T Extinction
7. Permian Extinction: 250 mya; K-T Extinction: 65 mya
8. 6
9. Cambrian Period; Ordovician Period; Silurian Period; Devonian Period; Carboniferous Period; Permian Period.

10. Cambrian Period: 542 mya; Ordovician Period: 488 mya; Silurian Period: 444 mya; Devonian Period: 416 mya; Carboniferous Period: 360 mya; Permian Period: 300 mya.

Review 12.2

1. b
2. a
3. d
4. d
5. b
6. d
7. b
8. d
9. a
10. c

Review 12.3

1. c
2. b
3. a
4. b
5. d
6. a
7. d
8. c

9. b
10. a

Review 12.4

1. d
2. a
3. d
4. c
5. b
6. a
7. b
8. d
9. a
10. b

Review 12.5

1. d
2. c
3. a
4. a, b, c, d
5. c
6. b
7. a
8. b
9. c
10. d

References

Introduction to Mnemonics

Numbered citations:

1. Soler, M.J. & Ruiz, J.C. 1996. The spontaneous use of memory aids at different educational levels. Applied Cognitive Psychology, 10, 41-51.

2. Xu Cui, Jeter, C.B., Yang, D., Montague, P.R. & Eagleman, D.M. 2007. Vividness of mental imagery: Individual variability can be measured objectively. Vision Research, 47, 474-8.

Other references:

Bellezza, F.S. 1981. Mnemonic Devices: Classification, Characteristics, and Criteria. Review of Educational Research, 51 (2), 247-275.

Bellezza, F.S. 1983. Mnemonic-device instruction with adults. In Pressley, M. & Levin, J.R. (eds.) Cognitive strategy research: Psychological foundations. New York: SpringerVerlag.

Bower, G.H. 1972. Mental imagery and associative learning. In L.W. Gregg (ed.) Cognition in learning and memory. New York: Wiley.

Higbee, K.L. 1997. Novices, Apprentices, and Mnemonists: Acquiring Expertise with the Phonetic Mnemonic. Applied Cognitive Psychology, 11, 147-161.

Morris, P.E. 1978. Sense and nonsense in traditional mnemonics. In M.M. Gruneberg, P.E. Morris & R.N. Sykes (eds.) Practical aspects of memory. London: Academic Press.

First-letter mnemonics

Numbered citations:

1. Boltwood, C.E. & Blick, K.A. 1970. The delineation and application of three mnemonic techniques. Psychonomic Science, 20, 339-341.

2. Nelson, D.L. & Archer, C.S. 1972. The first letter mnemonic. Journal of Educational Psychology, 63(5), 482-486.

3. Morris, P.E. & Cook, N. 1978. When do first letter mnemonics aid recall? British Journal of Educational Psychology, 48, 22-28.

4. Sergeant, A. & Gruneberg, M. 1997. The effectiveness of the first letter retrieval strategy in episodic memory. SARMAC II: Toronto, July 1997

5. Haring, M.J. & Fry, M.A. 1980. Facilitating prose recall with externally-produced mnemonics. Journal of Instructional Psychology, 7, 147-152.

Other references:

Carlson, L., Zimmer, J.W. & Glover, J.A. 1981. First-letter mnemonics: DAM (Don't Aid Memory). The Journal of General Psychology, 104, 287-292.

Waite, C.J., Blick, K.A. & Boltwood, C.E. 1971. Prior use of the first letter technique. Psychological Reports, 29, 630.

Rhythm & Rhyme

Numbered citations:

1. Schmuckler, M.A. 1997. Expectancy Effects in Memory for Melodies. Canadian Journal of Experimental Psychology, 51(4), 292-306.

2. Rainey, D.W. & Larsen, J.D. 2002. The Effect of Familiar Melodies on Initial Learning and Long-term Memory for Unconnected Text. Music Perception, 20 (2), 173-186.

3. Kimmel, K.J. 1998. The Development and Evaluation of a Music Mnemonic-Enhanced Multimedia Computer-Aided Science Instructional Module. Dissertation submitted to the Faculty of the Virginia Polytechnic Institute and State University in partial fulfillment of the requirements for the degree of Doctor of Philosophy in Teaching and Learning.

4. Cysarz, D., Von Bonin, D., Lackner, H., Moser, M. & Bettermann, H. 2004. Oscillations of heart rate and respiration synchronize during poetry recitation. American Journal of Physiology - Heart and Circulatory Physiology, 287(2), H587.

5. Bernardi, L., Sleight, P., Bandinelli, G., Cencetti, S., Fattorini, L., Wdowczyc-Szulc, J. & Lagi, A. 2001. Effect of rosary prayer and yoga mantras on autonomic cardiovascular rhythms: comparative study. BMJ, 323, 1446-1449.

Other references:

Wallace, W. T. (1994). Memory for music: Effect of melody on recall of text. Journal of Experimental Psychology: Learning, Memory, & Cognition, 20, 1471-1485.

Yalch, R. F. (1991). Memory in a jingle jungle: Music as a mnemonic device in communicating advertising slogans. Journal of Applied Psychology, 76, 268-275.

Keyword method

Numbered citations:

1. Hall, J.W., Wilson, K.P. & Patterson, R.J. 1981. Mnemotechnics: Some limitations of the mnemonic keyword

method for the study of foreign language vocabulary. Journal of Educational Psychology, 73, 345-57.

2. Rodríguez, M. & Sadoski, M. 2000. Effects of Rote, Context, Keyword, and Context/Keyword Methods on Retention of Vocabulary in EFL Classrooms. Language Learning, 50 (2), 385-412.

Brown, T.S. & Perry, F.L. Jr. 1991. A Comparison of Three Learning Strategies for ESL Vocabulary Acquisition. TESOL Quarterly, 25 (4), 655-670.

3. Andreoff, G.R. & Yarmey, A.D. 1976. Bizarre imagery and associative learning: A confirmation. Perceptual and Motor Skills, 43, 143-148.

Bergfeld, V.A., Choate, L.S. & Kroll, N.E. 1982. The effect of bizarre imagery on memory as a function of delay: Reconfirmation of the interaction effect. Journal of Mental Imagery, 6, 141-158.

Kroll, N.E.A., Schepeler, E.M. & Angin, K.T. 1986. Bizarre imagery: The misremembered mnemonic. Journal of Experimental Psychology: Learning, Memory and Cognition, 12, 42-53.

O'Brien, E.J. & Wolford, C.L.R. 1982. Effect of delay in testing on retention of plausible versus bizarre mental images. Journal of Experimental Psychology: Learning, Memory and Cognition, 8, 148-152.

Riefer, D.M. & Rouder, J.N. 1992. A multinomial modeling analysis of the mnemonic benefits of bizarre imagery. Memory & Cognition, 20, 601-611.

Webber, S.M. & Marshall, P.H. 1978. Bizarreness effects in imagery as a function of processing level and delay. Journal of Mental Imagery, 2, 291-300.

Campos, A., Amor, A. & González, M.A. 2002. Presentation of keywords by means of interactive drawings. The Spanish Journal of Psychology, 5(2), 102-109.

Campos, A. & Pérez, M.J. 1997. Mnemonic images and associated pair recall. Journal of Mental Imagery, 21, 73-82.

Kroll, N.E.A., Jaeger, G. & Dornfest, R. 1992. Metamemory for the bizarre. Journal of Mental Imagery, 16, 173-190.

Pra Baldi, A., De Beni, R., Cornoldi, C. & Cavedon, A. 1985. Some conditions for the occurrence of the bizarreness effect in free recall. British Journal of Psychology, 76, 427-436.

Riefer, D.M. & & LeMay, M.L. 1998. Memory for common and bizarre stimuli: A storage-retrieval analysis. Psychonomic Bulletin & Review, 5, 312-317.

Wollen, K.A. & Cox, S.D. 1981a. The bizarreness effect in a multitrial intentional learning task. Bulletin of the Psychonomic Society, 18, 296-298.

Wollen, K.A. & Cox, S.D. 1981b. Sentence cuing and the effectiveness of bizarre imagery. Journal of Experimental Psychology: Human Learning and Memory, 7, 386-392.

Marchal, A. & Nicolas, S. 2000. Is the picture-bizarreness effect a generation effect? Psychological Reports, 87, 331-340.

McDaniel, M.A., DeLosh, E.L. & Merritt, P.S. 2000. Order information and retrieval distinctiveness: Recall of common versus bizarre material. Journal of Experimental Psychology: Learning, Memory and Cognition, 26, 1045-1056.

Tess, D.E., Hutchinson, R.L., Treloar, J.H. & Jenkins, C.M. 1999. Bizarre imagery and distinctiveness: Implications for the classroom. Journal of Mental Imagery, 23, 153-170.

4. Beaton, A. A ., Gruneberg, M. M., Hyde, C. Shufflebottom, A. & Sykes, R.N. (2005). Facilitation of receptive and productive

foreign vocabulary acquisition using the keyword method: The role of image quality. Memory, 13, 458-471.

5. Raugh, M.R., Schupbach, R.D. & Atkinson, R.C. 1977. Teaching a large Russian language vocabulary by the mnemonic keyword method. U.S. Office of Naval Research, Technical Report 256.

6. Raugh, M.R. & Atkinson, R.C. 1975. A mnemonic method for learning a second-language vocabulary. Journal of Educational Psychology, 67, 1-16.

7. Wang, A.Y. & Thomas, M.H. 1992. The Effect of Imagery-Based Mnemonics on the Long-Term Retention of Chinese Characters. Language Learning, 42 (3), 359-376.

Wang, A.Y. et al. 1989. Do Mnemonic Devices Lessen Forgetting? Paper presented at the Annual Meeting of the American Psychological Association (97th, New Orleans, LA, August 11-15, 1989).

Wang, A.Y. et al. 1992. Keyword Mnemonic and Retention of Second-Language Vocabulary Words. Journal of Educational Psychology, 84, 520-8.

Wang, A.Y. & Thomas, M.H. 1995. Effect of keywords on long-term retention: help or hindrance? Journal of Educational Psychology, 87, 468-75.

Gruneberg, M.M. 1998. A commentary on criticism of the keyword method of learning foreign languages. Applied Cognitive Psychology, 12, 529-532.

Wang, A.Y. & Thomas, M.H. 1999. In defence of keyword experiments: a reply to Gruneberg's commentary. Applied Cognitive Psychology, 13, 283-287.

8. For example:

Hall, J.W., Wilson, K.P. & Patterson, R.J. 1981. Mnemotechnics:

Some limitations of the mnemonic keyword method for the study of foreign language vocabulary. Journal of Educational Psychology, 73, 345-357.

King-Sears, M.E., Mercer, C.D. & Sindelar, P.T. 1992. Toward independence with keyword mnemonics: A strategy for science vocabulary instruction. Remedial and Special Education, 13, 22-333.

9. Pressley, M., Levin, J.R., Digdon, N., Bryant, S.L. & Ray, K. 1983. Does method of item presentation affect keyword method effectiveness? Journal of Educational Psychology, 75, 686-91.

10. Johnson, R.E. 1974. Abstractive processes in the remembering of prose. Journal of Educational Psychology, 66, 772-9.

Jones, M.S., Levin, M.E., Levin, J.R. & Beitzel, B.D. 2000. Can vocabulary-learning strategies and pair-learning formats be profitably combined? Journal of Educational Psychology, 92, 256-62.

McDaniel, M.A. & Pressley, M. 1984. Putting the keyword method in context. Journal of Educational Psychology, 76, 598-609.

McDaniel, M.A., Pressley, M. & Dunay, P.K. 1987. Long-term retention of vocabulary after keyword and context learning. Journal of Educational Psychology, 79, 87-9.

Pressley, M., Levin, J. & Miller, G. 1982. The keyword method compared to alternative vocabulary-learning strategies. Contemporary Educational Psychology, 7, 213-26.

Shing, Y.S. & Heyworth, R.M. 1992. Teaching English Vocabulary to Cantonese-speaking Students with the Keyword Method. Education Journal, 20, 113-129.

11. Desrochers, A., Gélinas & Wieland, L.D. 1989. An

application of the mnemonic keyword method to the acquisition of German nouns and their grammatical gender. Journal of Educational Psychology, 81, 25-32.

12. Chongde, L., Tsingan, L. & Hongyu, L. 2004. The effect of mnemonic key-letters method on Chinese children at risk in English vocabulary learning. Acta Psychologica Sinica, 36 (4), 482-490.

Other references:

Atkinson, R.C. 1975. Mnemotechnics in second-language learning. American Psychologist, 821-8.

Atkinson, R.C. & Raugh, M.R. 1975. An application of the mnemonic keyword method to the acquisition of a Russian vocabulary. Journal of Experimental Psychology: Human Learning and Memory, 104, 126-133.

Bird, S.A. & Jacobs, G.M. 1999. An Examination of the Keyword Method: How Effective Is It for Native Speakers of Chinese Learning English? Asian Journal of English Language Teaching, 9, 75-97.

Campos, A., Amor, A. & González, M.A. 2004. The importance of the keyword generation method in keyword mnemonics. Experimental Psychology, 51(2), 1-7.

Campos, A., González, M.A. & Amor, A. 2003. Limitations of the mnemonic-keyword method. Journal of General Psychology, 130(4), 399-413.

Carney, R.N. & Levin, J.R. 1994. Combining Mnemonic Strategies to Remember Who Painted What When. Contemporary Educational Psychology, 19, 323-339.

Carney, R.N., Levin, J.R. & Stackhouse, T.L. 1997. The Face-Name Mnemonic Strategy from a Different Perspective. Contemporary Educational Psychology, 22 (3), 399-412.

Guey, C-c., Cheng, Y-y. & Huang, L-j. 2003. Effect of key-word method on memory of word groups for Chinese learners of English. Paper presented at the Hawaii International Conference on Education, Jan 7-10, Honolulu.

Hall, J.W. & Fuson, K.C. 1988. The keyword method and presentation rates: Reply to Pressley (1987). Journal of Educational Psychology, 80(2), 251-252.

Hall, J.W. 1988. On the utility of the keyword mnemonic for vocabulary learning. Journal of Educational Psychology, 80(4), 554-562.

Hall, J.W. 1991. More on the utility of the keyword method. Journal of Educational Psychology, 83(1), 171-172.

Jitendra, A.K., Edwards, L.L., Sacks, G. & Jacobson, L.A. 2004. What research says about vocabulary instruction for students with learning disabilities. Exceptional Children, 70(3), 299-322.

Levin, M. & Levin, J. 1990. Scientific mnemonomies: Methods for maximizing more than memory. American Educational Research Journal, 27(2), 301-321.

McDaniel, M.A. & Pressley, M. 1989. Keyword and context instruction of new vocabulary meanings: Effects on text comprehension and memory. Journal of Educational Psychology, 81(2), 204-213.

Pressley, M. 1991. Comparing Hall (1988) with related research on elaborative mnemonics. Journal of Educational Psychology, 83(1), 165-170.

Pressley, M., Levin, J.R., Nakamura, G.V., Hope, D.J., Bispo, J.G. & Toye, A.R. 1980. The Keyword Method of Foreign Vocabulary Learning: An Investigation of Its Generalizability. Journal of Applied Psychology, 65(6), 635-642.

Rosenheck, M.B., Levin, M.E. & Levin, J.R. 1989. Learning

botany concepts mnemonically: seeing the forest and the trees. Journal of Educational Psychology, 81, 196-203.

Thomas, M.H. & Wang, A.Y. 1996. Learning by the keyword mnemonic: Looking for long-term benefits. Journal of Experimental Psychology: Applied, 2(4), 330-342.

Uberti, H.Z., Scruggs, T.E. & Mastropieri, M.A. 2003. Keywords make the difference! Mnemonic instruction in inclusive classrooms. Teaching Exceptional Children, 35(3), 56-61.

Wang, A.Y., Thomas, M.H. & Ouellette, J.A. 1992. Keyword mnemonic and retention of second-language vocabulary words. Journal of Educational Psychology, 84(4), 520-528.

Willoughby, T., Wood, E. & Khan, M. 1994. Isolating variables that impact on or detract from the effectiveness of elaboration strategies. Journal of Educational Psychology, 86, 279-289.

Yeung, S.S. & Heyworth, R.M. 1992. Teaching English Vocabulary to Cantonese-speaking Students with the Keyword Method. Education Journal, 20(2), 113-129.

Extensions to the keyword method

Numbered citations:

1. Levin, J.R., Shriberg, L.K., Miller, G.E., McCormack, C.B. & Levin, B.B. 1980. The keyword method in the classroom: How to remember the states and their capitals. The Elementary School Journal, 82, 185-91.

2. For example,

McCormick, C.B. & Levin, J.R. 1984. A comparison of different prose-learning variations of the mnemonic keyword method. American Education Research Journal, 21, 379-398.

Peters, E.E. & Levin, J.R. 1986. Effects of a mnemonic imagery strategy on good and poor readers' prose recall. Reading Research Quarterly, 21, 179-192.

Shriberg, L.K., Levin, J.R., McCormick, C.B. & Pressley, M. 1982. Learning about "famous" people via the mnemonic keyword method. Journal of Educational Psychology, 74, 238-247.

3. McCormick, C.B., Levin, J.R., Cykowski, F. & Danilovics, P. 1984. Mnemonic-strategy reduction of prose-learning interference. Educational Communication and Technology Journal, 32(3), 145-152.

4. McCormick, C.B. & Levin, J.R. 1984. A comparison of different prose-learning variations of the mnemonic keyword method. American Education Research Journal, 21, 379-398.

5. Rummel, N., Levin, J.R. & Woodward, M.M. 2003. Do pictorial mnemonic text learning aids give students something worth writing about? Journal of Educational Psychology, 95(2), 327-334.

6. For example,

Carney, R.N. & Levin, J.R. 2000. Fading Mnemonic Memories: Here's Looking Anew, Again! Contemporary Educational Psychology, 25 (4), 499-508.

Carney, R.N. & Levin, J.R. 2000. Mnemonic instruction, with a focus on transfer. Journal of Educational Psychology, 92(4), 783-790.

7. Carney, R.N. & Levin, J.R. 2001. Remembering the Names of Unfamiliar Animals: Keywords as Keys to Their Kingdom. Applied Cognitive Psychology, 15, 133-143.

8. Carney, R.N. & Levin, J.R. 2003. Promoting higher-order learning benefits by Carney, R.N. & Levin, J.R. 2003. Promoting

higher-order learning benefits bybuilding lower-order mnemonic connections. Applied Cognitive Psychology, 17, 563-575.

9. Morrison, C.R. & Levin, J.R. 1987. Degree of mnemonic support and students' acquisition of science facts. Educational Communication and Technology Journal, 35(2), 67-74.

Story method

1. Bower, G.H. & Clark, M.C. 1969. Narrative stories as mediators for serial learning. Psychonomic Science, 14, 181-182.

2. Bower, G.H. 1973. How to ... uh... remember. Psychology Today, 7, 63-69.

3. Bellezza, F.S. 1982. Updating memory using mnemonic devices. Cognitive Psychology, 14, 301-27.

4. Cornoldi, C. & De Beni, R. 1996. Mnemonics and metacognition. In Herrmann, D., McEvoy, C., Hertzog, C., Hertel, P. & Johnson, M.K. (eds). Basic and Applied Memory Research, Vol 2: Practical Applications, 237-253.

5. Hishitani, S. 1985. Coding strategies and imagery differences in memory. Japanese Psychological Research, 27, 154-162.

6. Boltwood, C.E. & Blick, K.A. 1970. The delineation and application of three mnemonic techniques. Psychonomic Science, 20, 339-341.

Other references:

Cook, N.M. 1989. The applicability of verbal mnemonics for different populations: a review. Applied Cognitive Psychology, 3, 3-22.

Delaney, P.F. & Knowles, M.E. 2005. Encoding Strategy

Changes and Spacing Effects in the Free Recall of Unmixed Lists. Journal of Memory and Language, 52 (1), 120-130.

Drevenstedt, J. & Bellezza, F.S. 1993. Memory for self-generated narration in the elderly. Psychology and Aging, 8(2), 187-196.

Hill, R.D., Allen, C. & McWhorter, P. 1991. Stories as a mnemonic aid for older learners. Psychology and Aging, 6(3), 484-486.

Place method

Numbered citations:

1. DeBeni, R. & Cornoldi, C. 1985. Effects of the mnemotechnique of loci in the memorization of concrete words. Acta Psychologica, 60, 11-24.

2. Moè, A. & De Beni, R. 2004. Stressing the efficacy of the Loci method: oral presentation and the subject-generation of the Loci pathway with expository passages. Applied Cognitive Psychology, 19 (1), 95-106.

Other references:

Bellezza, F.S. 1983. The spatial-arrangement mnemonic. Journal of Educational Psychology, 75 (6), 830-837.

Gruneberg, M.M. 1992. The practical application of memory aids: Knowing how, knowing when, and knowing when not. In M.M. Gruneberg & P. Morris (eds), Aspects of memory, vol.1, London: Routledge, 2nd ed, pp 169-195.

De Beni, R., Moè, A. & Cornoldi, C. 1997. Learning from Texts or Lectures: Loci Mnemonics can Interfere with Reading but not with Listening. European Journal of Cognitive Psychology, 9 (4), 401-416.

Morris, P.E. 1979. Strategies for learning and recall. In M.M. Gruneberg & P.E. Morris (eds). Applied Problems in Memory. London: Academic Press.

Pegword Method

Numbered citations:

1. Morris, P.E. & Reid, R.L. 1970. Repeated use of mnemonic imagery. Psychonomic Science, 20, 337-338.

2. Carney, R.N. & Levin, J.R. 2010. Delayed mnemonic benefits for a combined pegword-keyword strategy, time after time, rhyme after rhyme. Applied Cognitive Psychology, In press.

3. Paivio, A. 1971. Imagery and verbal processes. New York: Holt, Rinehart and Winston.

Santa, J.L., Ruskin, A.D. & Yio, A.J.H. 1973. Mnemonic systems in free recall. Psychological Reports, 32, 1163-1170.

4. Delprato, D.J. & Baker, E.J. 1974. Concreteness of pegwords in two mnemonic systems. Journal of Experimental Psychology, 102, 520-522.

5. DiVesta, F.J. & Sunshine, P.M. 1974. The retrieval of abstract and concrete materials as functions of imagery, mediation and mnemonic aids. Memory and Cognition, 2, 340-344.

Link method

1. Bugelski, B.R. 1974. The image as mediator in one-trial paired-associate learning. III Sequential functions in serial lists. Journal of Experimental Psychology, 103, 298-303.

Delin, P.S. 1969. The learning to criterion of a serial list with and without mnemonic instructions. Psychonomic Science, 16, 169-170.

Coding mnemonic

Numbered citations:

1. Chase, W.G. & Ericsson, K.A. 1981. Skilled memory. In J.R. Anderson (ed.) Cognitive skills and their acquisition. Hillsdale, NJ: Erlbaum.

2. Higbee, K.L. 1997. Novices, Apprentices, and Mnemonists: Acquiring Expertise with the Phonetic Mnemonic. Applied Cognitive Psychology, 11, 147-161.

3. Patton, G.W.R. & Lantzy, P.D. 1987. Testing the limits of the phonetic mnemonic system. Applied Cognitive Psychology, 1, 263-71.

Patton, G.W.R., D'Agaro, W.R. & Gaudette, M.D. 1991. The effect of subject- generated and experimenter-supplied code words on the phonetic mnemonic system. Applied Cognitive Psychology, 5, 135-48.

4. Kliegl, R., Smith, J., Heckhausen, J. & Baltes, P.B. 1987. Mnemonic Training for the Acquisition of Skilled Digit Memory. Cognition and Instruction, 4 (4), 203-223.

5. Bellezza, F.S., Six, L.S. & Phillips, D.S. 1992. A mnemonic for remembering long strings of digits. Bulletin of the Psychonomic Society, 30 (4), 271-274.

6. Hatano, G. & Kuhara, K. 1973. Production and use of mnemonic phrases in paired-associate learning with digits as response terms. Psychological Reports, 33, 923-930.

7. Higbee, K.L. 1988. You Memory: How it works and how to improve it. NY: Simon & Schuster, Inc.

Other references:

Morris, P.E. & Greer, P.J. 1984. The effectiveness of the phonetic mnemonic system. Human Learning, 3, 137-142.

Slak, S. 1970. Phonemic recoding of digital information. Journal of Experimental Psychology, 86, 398-406.

Slak, S. 1985. On phonetic and phonemic mnemonic systems: A reply to M.J. Dickel. Perceptual and Motor Skills, 61, 727-733.

Wilding, J. & Valentine, E. 1994. Mnemonic Wizardry with the Telephone Directory — But Stories are Another Story. British Journal of Psychology, 85, 501-509.

Mastering mnemonics

Numbered citations:

1. Baddeley, A.D. & Lieberman, K. 1980. Spatial working memory. In Nickerson, R.S. (ed.), Attention and Performance VIII: Hillsdale NJ: Erlbaum.

Logie, R.H. 1986. Visuo-spatial processing in working memory. Quarterly Journal of Experimental Psychology, 38A, 229-47.

2. Brewer, W.F. 1980. Literary theory, rhetoric, and stylistics: Implications for psychology. In R. J. Shapiro, B. C. Bruce, & W. F. Brewer (Eds.), Theoretical issues in reading comprehension (pp. 221-239). Hillsdale, NJ: Erlbaum.

3. Bower, G.H. & Reitman, J.S. 1972. Mnemonic elaboration in multilist learning. Journal of Verbal Learning and Verbal Behavior, 11, 478-485.

4. Solso, R. L. 1995. Cognitive psychology (4th ed.). Boston: Allyn & Bacon. 5. Wilding, J. and Valentine, E. 1994. Memory champions. British Journal of Psychology, 85, 231-244.

6. Cepeda, N.J. et al. 2008. Spacing Effects in Learning: A Temporal Ridgeline of Optimal Retention. Psychological Science, 19 (11), 1095-1102.

7. Taylor, K. & Rohrer, D. 2009. The effects of interleaved practice. Applied Cognitive Psychology, Published online 30 July.

8. Bjork, R. A. 1994. Memory and metamemory considerations in the training of human beings. In J. Metcalfe, A. Shimamura, (Eds.), Metacognition: Knowing about knowing (pp. 185-205). Cambridge, MA: MIT Press.

Other references:

Higbee, K.L. 1988. You Memory: How it works and how to improve it. NY: Simon & Schuster, Inc.

English-Spanish Glossary

Acronyms & acrostics can be found at the end

abducted — secuestrado

abducting — secuestrando

abilities — habilidades

absolutely — absolutamente

absorbed — absorbido

abstract — abstracto, se refiere a conceptos no físicos, como la justicia o el amor

abundant — abundante

accessory — accesorio

accounts — cuenta

acoustically — acústicamente

acquiring — adquisidor

acronyms — acrónimos

actions — acciones

active — activo

actual — real, verdadero

addendums — apéndices

adequate — adecuado

adjuncts — adjuntos, usado en conjunto con otras estrategias

advantages — ventajas

advised — aconsejado

advocating — abogando, defendiendo

albeit — aunque
algal bloom — floración de algas
anchors — anclas
ancient — antiguo
anecdotal — anecdótico
anonymous — anónimo
antiquity — antigüedad
apparent — evidente
applicable — aplicable, capaz de aplicar
applying — aplicando
appropriate — apropiado
approximately — aproximadamente
aquatic — acuático
arbitrary — arbitrario
art appreciation — apreciación artística
articulating — articulando, expresando claramente
artificial memory aids — ayudas artificiales de memoria
artist's style — estilo del artista
artwork — obra de arte
aspect — aspecto
assistance — asistencia, ayuda
associated — asociado
associations — asociaciones
assuredly — ciertamente
astray — extaviado

at the expense of — a expensas de

attaching — adjuntando

attention — atención

attest — dar fe, testificar

attributes — atributos, caracteristicas

attribution error — error de atribución, la tendencia a creer que lo que hace la gente refleja lo que son

auditory — auditivo

automatically — automáticamente

autumn — otoño

awkwardness — torpeza

background knowledge — conocimiento de fondo

banality — banalidad

bare-bones — lo esencial

barking — ladrido

basic principles — principios básicos

basilisk — basilisco

bear in mind — tener en cuenta

bed — cama

belief — creencia

benefits — beneficios

binds — enlaza

biographical — biográfico

biographies — biografías

birdwatcher — observador de aves

bizarre — extraño

blindfolded — con los ojos vendados

blood pressure — presión sanguínea

bogged down — empantanado

bonnet — capó

boring — aburrido

borne in mind — tener en cuenta

bother — te molestes

boundary — límite

brevity — brevedad

brilliant — brillante, excelente

brute force — fuerza bruta

burden — carga

by repute — por su reputación

by virtue of — en virtud de

candidates — candidatos

captioned — subtitulado

categories — categorías

ceiling — techo

cellular physiology — fisiología celular

cement — cimentamos, consolidamos

characteristic — característica

chilly — frío

chunking — haciendo trozos

chunks — trozos

circulation patterns — patrones de circulación

circumstances — circunstancias

cited — citado

clapping — aplaudiendo

classical epics — epopeyas clásicas

classifications — clasificaciones

closet — armario

clown — payaso

clubhouse — casa club

clues — pistas

clustering — recogiendo en grupos significativos

clusters — racimos, grupos

cluttering — abarrotando

coal — carbón

cognate — cognado

coherent — coherente, lógico

collided — colisionado

combining — combinando

comfortable — cómodo

comma — coma

commemorating — conmemorando

common — común, ordinario

comparison — comparación

compelling — irresistible

complex — complejo, complicado

complicated — complicado

composite — compuesto

comprehension — comprensión

computer animated — animado por computadora

concentrate — concentres

concepts — conceptos

conclusions — conclusiones, resultados

concrete — concreto, se refiere a cosas físicas

concreteness — concreción

confess — confieso, admito

confirm — confirman

conflicting — contradictorio

confusing — confuso

conjunction — en conjunto, juntos

connections — conexiones

connotations — connotaciones

conscious awareness — conocimiento consciente

consecutive — consecutivo, siguiendo en secuencia

consequent — consiguientes, resultantes

considered — considerado

consistency — consistencia

consolidation — consolidación

consonants — consonantes

constraint — restricción

construct — construir

contentious — contencioso

context — contexto

contextual — apropiado al contexto habitual

contributions — contribuciones

contrived — artificial, forzado

control condition — condición de control, grupo que se usa como comparación base

conversation — conversacion

convict — convicto, presidiario

convincingly — convincentemente

correctly — correctamente

correlated — correlacionado

correspond — corresponden

costliest — más costoso

couch — sofá

counterpart — contraparte

counters — contrarresta

crafting — elaboración

cram for an exam — prepararse apresuradamente

cravat — corbata

creation — creación

critical — crítico

criticism — crítica

criticized — criticado

crutch — muleta

cues — señales

cultural — cultural

currant — pasa de Corinto

dactylic hexameter — hexámetro dactilico, donde cada pie tiene una sílaba larga y dos cortas

daylight saving — hora de verano

dazzling — deslumbrante

decidedly — decididamente

deciding issue — cuestión decisiva

decoding — descodificación

defining event — evento que define una era

definition — definición

degree — grado

deliberate attempt — intento deliberado

demands — demandas, necesidades

demonstrated — demostrado

deportment — porte

derailed — descarriló

derive — derivar

derived — derivado

descriptive — descriptivo, texto que describe a una persona, lugar o cosa

designed — diseñado

desirable — deseables, aconsejables

desk — escritorio

destroys — destruye
detailed — detallado
determining — determinando
devise — idear, inventar
dice — dados
different — diferente
diligently — diligentemente
dimensions — dimensiones
dinosaurs — dinosaurios
direction — dirección
disadvantages — desventajas
disappear — desaparecer
discipline — disciplina, sujeto
disconsolate — desconsolado
discussions — discusiones
disentangled — desvinculados
dismember — desmembrar
disrupted — interrumpido
dissimilar — disímil, diferente
distinction — distinción
distinctiveness — diferencia, carácter distintivo
distinguish — distinguir
distract — distraer
distributing — distribución
divorce — divorcio

do away with — eliminar
double burden — doble carga
downstroke — trazo que baja
dramatic — dramático, grandes
drawbacks — desventajas
drifting — deriva
dry — seco
drying out — el secado
due — adecuado
durable — durable, duradero
echo — eco
educational — educativo
effectiveness — eficacia
effects — efectos
effort — esfuerzo
elaborated — elaborado
elements — elementos
embed — incrustar
embedded — se incrustado
embryo — embrión
emotion — emoción
emphasis — énfasis
emphasized — enfatizado
employs — emplea
enables — habilita

encapsulates — encapsula, capta

encoded — codificado

encompass — abarcan

endeavored — esforzado

energies — energías

engages — se compromete

engine — motor

enhance — mejorar

enough give — suficiente flexibilidad

entirety — totalidad

entwined — entrelazado

equipment — equipo

equivalent — equivalente, lo mismo

eraser — goma de borrar

essence — esencia

essentially given up — esencialmente abandonado

evaluating — evaluando

everyday task — tarea diaria

evoking — evocando

exam anxiety — ansiedad de examen

examinations — exámenes

examples — ejemplos

exchange — intercambio

executive function — función ejecutiva, procesos cognitivos que lo ayudan a controlar sus pensamientos y comportamiento

exhibition — exposición

expected — esperado

experiences — experiencias

expert database — base de datos de expertos

expertise — pericia

explicitly — explícitamente, declarado abiertamente

exposed — expuesto

expository text — texto expositivo, texto cuyo propósito es explicar o informar

extension — extensión

extensive — extensa

extinctions — extinciones

extinguished — extinguido

extraordinary — extraordinario

extreme — extremos

eyelash — pestaña

face-name association — asociación cara-nombre

facility — facilidad, talento

factors — factores

facts — hechos

fae — hada

faithful — fiel

feel-good — sentirse bien, agradable

fictional — ficticio

fidelity — fidelidad

figures — cifras

filed — archivado

finding — hallazgo

fits in nicely — se adapta bien

fixed — fijo

flat — plano

fleet — flota

flexibly — de una manera flexible

floated — flotó

floor — piso

fluently — fluido

focus — atención

fool — tonto

foremost — principal, primero

forgotten — olvidado

formalized — formalizado, dada una forma definida

former — primero

frame — marco

framework — marco

frequent — frecuente

full stop — punto

function — función

further — adicional

fuzzy — borroso

gallows — horca

gathered — reunido

gem — joya

gender — género

generating — generando

genesis — génesis, origen

geological time scale — escala de tiempo geológica

giant — gigante

glaringly — claramente obvio

glossy photograph — fotografía brillante

gospel — evangelio

graphic — gráfico

grasslands — pastizales

gross — brutas

grounded — se basa

guidelines — directrices

guitar — guitarra

habitually — por costumbre

hangs out of — cuelga de

hesitation — vacilación

hexametric — hexamétrico, tiene una línea de versos de seis pies métricos

hierarchical — jerárquico

highlighting — destacando

highlights — destaca

highly prized — muy apreciado

hiking — excursionismo

hinder — impedir

historical — histórico

holistic — holístico, tratado como un todo

host of — gran cantidad de

hypocritical — hipócrita

hypodermic — hipodérmico

identifying — identificando

ignored — ignorado

illustrating — ilustrando

illustration — ilustración, imagen

images — imágenes mentales

imageability — capacidad de imagen

imagery — imágenes

imagined — imaginado

immediately — inmediatamente

impacts — impacta

implement — poner en práctica

implied — impliqué

impress — impresionar

impressive — impresionante

impulse — impulso

in a similar vein — en un sentido similar

in broad strokes — en líneas generales

in depth — a fondo

in mind — en mente

in strict accordance — en estricto acuerdo

incidental — no dirigido

inconsistent — inconsistente

incorporating — incorporando

indictment — acusación

individual variability — variabilidad individual

inexplicable — inexplicable, no se puede explicar

information — información

inherently — intrínsecamente

initial — inicial

insidiously — insidiosamente

instructed — instruido

instructional module — módulo de instrucción

integrated — integrado

intelligible — inteligible

intensive — intensivo

interactive — interactivo

interference — interferencia

interleaving — intercalar

intermediary — intermediario, actúa como un enlace entre dos cosas

intervals — intervalos

interwoven — entretejido

intriguing — intrigante

intuitively — intuitivamente
invariably — invariablemente
invented — inventadas
issue — tema
it is crucial that — es crucial que
items — artículos
ivy — hiedra
jaw — mandíbula
jawless fish — pescado sin mandíbula
jester — bufón, payaso
jingle — rítmico sonsonete
journey — viaje
judge decrees — el juez decreta
justice — justicia
kelp forest — bosque de algas marinas
keyword — palabras clave
knowledgeable — bien informado
lacking — carente
lamp — lámpara
lamprey — lamprea
landbridge — puente de tierra
landmarks — hitos, puntos de referencia
lapel — solapa
latter — segundo
learning — aprendizaje

learning disabled — con una dificultad de aprendizaje

leash — correa

leaves — hojas

lecture — conferencia

lend itself — prestarse

lessens — disminuye

lesson plans — planes de lecciones

likewise — de la misma manera

limber — ágil

limitations — limitaciones

links — enlaces, conexiones

lists — listas

lizard — lagarto

logical — lógico

long-lasting — duradero

long-term memory — memoria a largo plazo

lumping together — concentrándose juntos

lungs — pulmones

lyrics — letra

magic bullet — bala mágica

magician — ilusionista

main points — puntos principales

major aspects — aspectos principales

majority — mayoría

making up — componiendo, inventando

mammals — mamíferos

markedly better — notablemente mejor

mass extinction — extinción masiva

massive — masivo

master — dominar

mastered — dominado

mastodon — mastodonte

match — corresponde a

mathematical operations — operaciones matemáticas

matrix — matriz

meaning — sentido

meaningful connections — conexiones significativas

meaningless — sin sentido

measure — medida

measurement — medición

mediators — mediadores, intermediarios

medical — de medicina

melody — melodía

memorized — memorizado

memory codes — códigos de memoria

memory improving strategy — estrategia para mejorar la memoria

mentioned — mencionado

meter — metro

method of loci — método de loci

microscopic creatures — criaturas microscópicas

microwave — microonda

mind go blank — mente en blanco

minerals — minerales

minimize — minimizar

mixed — mezclado

mnemonic strategy — estrategia mnemotécnica

mnemonist — mnemonista, un experto en técnicas mnemotécnicas

modeled — modelado

modified — modificado

motivation — motivación

mountain — montaña

movement — movimiento

muddy — fangoso

multiple — múltiple

mysterious — misterioso

narrow — estrecho

naturalistic — naturalista

neat — ordenado

necessarily — necesariamente

nested — anidado

network — red

noose — soga

not uncommonly — no es raro

notes — notas

noteworthy — digno de mención

noticeable — que se nota, evidente

notorious — notorio

notwithstanding — a pesar de

novices — novatos

numbered — numerado

nursery song — canción infantil

obliterate — obliterar, borrar

oblivious — inconsciente de

obvious — obvio

off the top of my head — sin pensarlo dos veces

old-fashioned — anticuado

on the fly — sobre la marcha, improvisado

on track — en camino de

opportunities — oportunidades

opposite — opuesto

optician — óptico

optimal — óptima

order — orden

ordered — ordenado

ordinary — ordinario

organisms — organismos

organizational — organizativo

originally — originalmente

out-performed — superado
over-learned — sobre-aprendido
overcome — superar
overlooked — pasado por alto
overwhelming — abrumador
owl — búho
oxygen — oxígeno
paintbrush — brocha
paradoxically — paradójicamente
parcel — paquete
parody — parodia
parrot — loro, papagayo
passage — pasaje de texto
paste on — pegar en
peculiarity — rareza
pedestrian mall — centro comercial peatonal
pegs — clavijas
pegword method — método della parola aggancio
pencil — lápiz
perceived — percibido
perched — posado
performance — ejecución
perils — peligros
periodic table — tabla periódica de los elementos
permanently — permanentemente

permissible — permisibles, admisibles
permitted — permitido
persevere — perseverar
persistence — persistencia
phantom — fantasma
phonetic — fonético, relacionado con el sonido de las letras
phonetically — fonéticamente
phrases — frases
physical — físico
pigment — pigmento
planets — planetas
plate tectonics — placas tectónicas
pleasurable — agradable
pliable — flexible
pliers — alicates
poems — poemas
pointless — inútil
pong — hedor
potentially — potencialmente
pottery — cerámica
power — poder
practice — práctica
pre-learned — pre-aprendido
precedence — precedencia, prioridad
precipitous drop — caída precipitada

precise — preciso, exacto
predictable — previsible
predominantly — predominantemente
preference — preferencia
presentation — presentación
presupposes — presupone
primacy — primacía
prior — anterior
problems — problemas
processe — procesarse
processing speed — velocidad de procesamiento
prominent — prominente, conspicuo
prone — propenso a
pronounceable — pronunciable
prose — prosa
protean — proteico, mudable
psychologist — psicólogo
puddles — charcos
puppet — marioneta
purpose — propósito
quality — calidad
reaction times — tiempos de reacción
readily — fácilmente
readily to mind — viene a la mente de inmediato
recalled — recordado

recap — recapitulemos

recency — actividad reciente

recitation — recitación

reciting — recitando

recognition — reconocimiento

recognize — reconocer

recordings — grabaciones

recovered — recuperado

recycling — reciclaje

reduce — reducen

redundant — redundantes

reference — referencia

refine — refinar, perfeccionar

reflects — refleja

regal — real

regalia — insignias reales

regardless — independientemente

regurgitate — regurgitar (repetir peatonalmente)

rehearsal — ensayo, repetición

reinforces — refuerza

reiterating — reiterando, repitiendo

related — relacionado

relational — relacional

relationships — relaciones

release — emisión

relevant — pertinente

reliably — de manera confiable

reliance — confianza

relishes — disfruta

remedy — remedio

remembered — recordado

reminders — recordatorios

reminiscent — recuerdan

rendered — traducido

reorderings — reordenamientos

repeat — repetir

repetition — repetición

replacement — reemplazo

represented — representado

research — investigación

resorted to — recurrido a

respectively — respectivamente

responsiveness — reactividad

restraints — restricciones

retrieval — recuperación

retrieve — recuperar

reversal — inversión

revolutionary — revolucionario

rhetoric — retórica

rhymes — rimas

rhythm — ritmo
ribs — costillas
rolls off the tongue — sale de la lengua
root — raíz
rote repetition — repetición mecánica
routes — caminos
rule-of-thumb — regla práctica, heurístico
ruler — regla
rut — surco
saber-tooth — tigre dientes de sable
scales — equilibrio
school uniform — uniforme escolar
scissors — tijeras
seeking — buscando
selected — seleccionado
selectively — selectivamente
semantic — semántico, relacionado al significado
seminar — seminario
sensitivity — sensibilidad
sentence — oración, frase
separate — separado
sequence — secuencia
serial position — posición serial
setting — entorno
setting the stage — preparando el escenario

settled — estable

sharps — sostenidos

shelter — abrigo

shields — escudos

shivering — tiritando

shortcut — atajo

sickle — hoz

significant difference — diferencia significativa

signify — significar, indicar

silver bullet — bala de plata

similar — similares, parecidos

simple — sencillo

simplicity — sencillez

simplified — simplificado

simultaneously — al mismo tiempo

singularly unhelpful — realmente muy inútil

situations — situaciones

skeleton — esqueleto

skill — técnica, habilidad

slate — pizarra

slight — leve

sludge — lodo

smile — sonrisa

snowball — bola de nieve

social studies — ciencias sociales

sock — calcetín
solaces — consola
soldiered through — tercamente empujado
sound bite — frase incisiva
sounding — suenan similares
spaced repetition — repetición espaciada
spaces — espacios
spatial — espacial
specialist topics — temas especializados
specifically — específicamente, en particular
specified — especificado
speculated — especulado
speech — habla
splashes — salpica
spuriously — engañosamente
squashy — suave
squeezing — exprimido
statistics — estadística
store — base de datos
stories — cuentos
straightforward — sencillo
strange — extraño
strategy — estrategia
strengthened — fortalecido
string — cadena, secuencia

structural coherence — coherencia estructural

structure — estructura

strumming — rasgueo

stuck — pegado

sub-topics — subtemas

subject — tema

submerged — sumergido

substitution — sustitución

subsume — subsumen, incluyen

succeeded by — seguido por

sucker — ventosa

suffix — sufijo

suggested — sugirió

summarize — resumir

summit — cumbre

sums — sumas, aritmética

superior — mayor

supper — cena

supplied — se proporcionó

support — apoyo

surf lifesaving — de salvamento de vidas de oleaje

surfeit — exceso

surmised — conjeturado

suspect — sospecha

swallow — golondrina

swelling — hinchazón

swigging — tragando

syllable — sílaba

symmetry — simetría

synchronized — sincronizado

system — sistema

tackled — abordado

talent — talento, dotes

tamped — apisonado

target — objetivo

tattooed — tatuado

taxonomy — taxonomía

technical words — términos técnicos

techniques — técnicas

teetering — tambaleante

tellingly — eficazmente

temporal order — orden temporal, organización de eventos en el tiempo

terribly — muy

texts — textos

theme — tema

theories — teorías

theorists — teóricos

thermodynamics — termodinámica

thoroughly — completamente

tidal wave — oleada, ola gigante

toolbox — caja de herramientas

torrent — torrente, inundación

traces — huellas

trade-offs — compromisos

traditional — tradicional, clásico

transformational — transformacional

transformed — transformado

transition — transición

trapped — atrapado

treble staff — pentagrama clave del sol

tribe — tribu

tricks — trucos

tricky — complicado

triggered — activado

trunked elephants — elefantes troncos

tune — melodía

tweaking — retocar, cambiar un poco

typically — típicamente

ultimately — por último

unconnected — desconectado

uncool — no genial

underscores — subraya

understandable — comprensible

underwater — submarino

unfamiliar — desconocido
unfortunately — desafortunadamente
uniformity — uniformidad
uninformative — eso no da información
unobservant — inobservante
unparalleled explosion — explosión sin precedentes
unrelated — no relacionado
unusual — raro
usefulness — utilidad
valuable — valioso
variant — variante, variación
variation — variación
verbal — verbal, para hacer con palabras
verse — poesía, versos
vertebrate — vertebrado
vexed — espinoso
victims — víctimas
visual cortex — corteza visual
visual scene — escena visual
visualize — visualizar, imaginar
visualizers — visualizadores
vividness — viveza
vocabulary — vocabulario
vocalizing — vocalizante, diciendo en voz alta
volunteers — voluntarios

vulture — buitre

wall — pared

wary — cauto

wasted effort — esfuerzo inútil

wastepaper bin — papelera

waterlily — lirio de agua

wavers — tambalea

waving — agitando

weaker — más débil

weird — extraño

well-versed — bien versado

whereby — por lo cual

wiping out — aniquilando

witch — bruja

woolly mammoth — mamut lanudo

word labels — etiquetas de palabras

working memory — memoria de trabajo

works wonders — funciona de maravilla

yard — patio

Acronyms & acrostics

BEDMAS:

Brackets — Soportes

Exponent — Exponente

Division — División

Multiplication — Multiplicación

Addition — Adición

Subtraction — Sustracción

Characteristics of living things:

Movement — Movimiento

Respiration — Respiración

Sensitivity — Sensibilidad

Growth — Crecimiento

Reproduction — Reproducción

Excretion — Excreción

Nutrition — Nutrición

cranial nerves — nervios craneales:

olfactory — olfatorio

optic — óptico

oculomotor — motor ocular comun

trochlear — troclear o patético

trigeminal — trigémino

abducens — abducens o ocular externo

facial — facial
auditory — auditivo
glossopharyngeal — glosofaríngeo
vagus — neumogástrico
accessory — accesorio o espinal
hypoglossal — hipogloso mayor

Planets:
Mercury — Mercurio
Venus — Venus
Earth — Tierra
Mars — Marte
Jupiter — Júpiter
Saturn — Saturno
Uranus — Urano
Neptune — Neptuno
Pluto — Plutón

Taxonomy of Living things:
Kingdom — Reino
Phylum — Filo
Class — Clase
Order — Orden
Family — Familia
Genus — Género
Species — Especie

www.ingramcontent.com/pod-product-compliance
Lightning Source LLC
Chambersburg PA
CBHW071558080526
44588CB00010B/945